CENTENNIAL FARMS of Indiana

IHS CENTENNIAL FARM FAMILIES RECORDS

Centennial Farm medallion

CENTENNIAL FARMS of Indiana

Edited by

M. Teresa Baer

Kathleen M. Breen

Judith Q. McMullen

With Genealogical Indexes by

Ruth Dorrel

INDIANA HISTORICAL SOCIETY PRESS

INDIANAPOLIS 2003

Printed in China

This book is a publication of the
Indiana Historical Society Press
450 West Ohio Street
Indianapolis, Indiana 46202-3269 USA
www.indianahistory.org
Telephone orders 1-800-447-1830
Fax orders 317-234-0562
Orders by E-mail shop.indianahistory.org

The paper in this publication meets the minimum requirements of American National Standard for Information Sciences—Permanence of Paper for Printed Library Materials, ANSI Z39.48–1984. ∞

Library of Congress Cataloging-in-Publication Data

Centennial farms of Indiana / edited by M. Teresa Baer, Kathleen M. Breen, Judith Q. McMullen ; with genealogical indexes by Ruth Dorrel.
p. cm.
Includes indexes.
ISBN 0-87195-164-9 (alk. paper)
1. Farm life—Indiana—History. 2. Agriculture—Indiana—History. 3. Farms—Indiana—History. 4. Indiana—Genealogy. 5. Indiana—History, Local. I. Baer, M. Teresa, 1956– II. Breen, Kathleen M. III. McMullen, Judith Q., 1945– IV. Dorrel, Ruth.

F526.C46 2003
630'.9772—dc21 12002032815

Contents

Shirley (Snyder) McCord
The Career of a Hoosier Historical Editor

Shirley (Snyder) McCord retired from the Indiana Historical Bureau on 27 September 1991 after thirty-one years of service to the state of Indiana as an historical editor. Her service was recognized with designation as a Sagamore of the Wabash by Gov. Evan Bayh.

Shirley worked for the Bureau from 1947 to 1952 under Howard H. Peckham and contributed to the production of several publications and projects, including the Centennial Farms Program of the Indiana Historical Society. She shared editorial credit with Peckham on the 1948 *Letters from Fighting Hoosiers*, a volume in the Indiana War History Commission project *Indiana in World War II*.

Shirley returned to the Indiana Historical Bureau in September 1965 under Hubert H. Hawkins and worked until 1991. She was a legend in the Indiana State Library and Historical Building and the historical community at large for her expertise in editing and her knowledge of Indiana history. During her tenure, she contributed her significant talents to the completion of at least twenty volumes for the Bureau, including the precedent-setting *The Centennial History of the Indiana General Assembly, 1816–1978.* She also worked with and aided many notable scholars and authors of Indiana history. On a daily basis, Shirley helped to make all the work of the Bureau better, watching over the periodicals and educational materials that continue today to enrich the lives of Hoosiers and those who study the Hoosier State.

Shirley's contributions to the preservation and dissemination of Indiana history did not end with her formal retirement. She worked on a part-time basis with the Publications Division of the Indiana Historical Society for nearly eight years. In that capacity—among other contributions—she helped to bring to completion volume 2 of *The History of Indiana* by Donald F. Carmony, emeritus professor of history, Indiana University Bloomington. Her knowledge of the period covered by the volume and her work with Carmony was crucial in bringing *Indiana, 1816–1850: The Pioneer Era* into print after a long and arduous process of research, writing, and editing.

Shirley (Snyder) McCord died 20 September 1999. Sadly, no future Indiana historians will benefit from her expertise. Anyone who studies the Hoosier State, however, reaps the benefit of her decades of dedication to Indiana history.

Pamela J. Bennett
Indiana Historical Bureau

INDIANA HISTORICAL BUREAU

Shirley (Snyder) McCord receiving a Sagamore of the Wabash Award in 1991

An Overview of the Centennial Farms Award Program

Judith Q. McMullen

Anything that lasts for one hundred years is remarkable. But when the centenarian is a family farm, it is even more worthy of attention. In 1947, in conjunction with Purdue University, the Indiana Historical Society began a program, originally called Pioneer Farms, to honor families who had owned a particular farm for one hundred years or longer. The Indiana Historical Society initiated the Centennial Farms program, stating "that this stable element in Hoosier society deserves some recognition."[1] The first announcement of the program in the January 1947 *Indiana History Bulletin* noted: "About forty-five per cent of the population of Indiana still is rural; farming is yet a big business in the state. In many counties generation after generation has stayed with the land, tilling the soil that was broken by grandfathers or great-grandfathers. . . . The Indiana Historical Society is going to find out how many farms have been in the same families one hundred years or more."[2]

In April 1947 Howard H. Peckham, secretary of the Indiana Historical Society, commented, "At the beginning of this project, your secretary blithely assumed that there might be 50 to 75 farmers whose land had been a family possession for a century or more." However, by the time of his report, the Society had "a roster of 400 farms."[3] By the year's end, the Society reported that "the total number of centennial farm families has mounted toward 750 and is still growing."[4] Peckham stated that farmers "are delighted that some state organization is giving them recognition for their long and steadfast devotion to the soil."[5]

At a meeting of the executive committee of the Society, members heard "a report on the astonishing number of farm families in Indiana," who were eligible for the award.[6] According to the report, "The project was not conceived in such proportions, but the co-operation of Purdue University's county agricultural agents has brought to light this large number of Indiana families who have stayed by the land of their ancestors."[7] Purdue University enlisted the help of the county agricultural agents in Indiana's ninety-two counties to locate centennial farms. A letter sent out to the agents from the secretary of the Indiana Historical Society and the dean of the College of Agriculture at Purdue explained the program: "A direct lineal descendant, man or woman, must still own the century-old farm, or a good portion of it. It is not necessary that the present owner be working the farm, or living on it—just so he owns the farm that his forefather acquired 100 years ago or more."[8]

Our farm has been in the same family for 102 years. . . . Our youngest son Fred D. Kunkel, is expecting to represent the fifth generation to live on this farm.

Mrs. Sherman Kunkel
Adams County

Opposite: Left: Form letter from Indiana Historical Society and Purdue University to Indiana county agricultural agents announcing the Centennial Farms Program. Right: Form letter from Indiana Historical Society to Centennial Farm families, which accompanied the certificates of recognition and announced the availability of bronze medallions.

IHS INDIANA EXTENSION HOMEMAKERS COLLECTION

The 1947 *Indiana History Bulletin* published a preliminary report on the locations by county of the initial 706 applications for Hoosier Centennial Farm status. The report speculated that "farms in rich agricultural areas tend to be held in old families and handed down from one generation to the next." This could certainly be said of most of the counties with the highest numbers of applications: Rush County, 54 applications; Wayne County, 46 applications; Johnson County, 37 applications; Jefferson County, 35 applications; and Grant County, 27 applications. The report also stated that more of the centennial farms "are found in the southern part of the state than in the northern because the former is older." Accordingly, 19 percent of the centennial farms were located in the thirteen counties that bordered the Ohio River.[9]

The relative enthusiasm of the county agricultural agents also appears to have affected the number of applications submitted from counties. While Peckham acknowledged that the agents had "been most helpful in locating many of these farms," some of the agents were extremely conscientious in seeking applicants.[10] For example, in Rush County, the county from which the largest number of applications was submitted, the agent actually filled out most of the forms for the applicants.

Once completed, the application forms were mailed to the Society's offices. At that time, the Indiana Historical Society and the Indiana Historical Bureau shared offices and staff. Shirley Snyder, a member of the Bureau's staff, was given the difficult task of checking the land records in the Indiana State Archives to prove or disprove the accuracy of the information given on the applications. The job required both patience and diplomacy. Snyder often had to write applicants for further information or to inform them that she could find no record of their farms. Although most applicants were eager to answer questions concerning their farm's eligibility, some got a bit testy. One applicant wrote, "Replying to your recent request for additional historical facts as to the length of time the FARM I STILL OWN has been in my family name, I can only say what I recall from what my father told me years ago." Another applicant responded, "It gives me great pleasure in informing you that your information is incorrect." And yet another answered Snyder's request for more substantial information by stating, "I am really the *only one* who knows the definite facts." One woman wrote emphatically that the Society had made a mistake in spelling her name, "PLEASE NOTICE . . . you MIS-SPELLED my family name. Your typist wrongfully omitted the 't' (fourth letter from the last letter in the word)."[11] Despite such outpourings from a few of the correspondents, Snyder's responses were always courteous. The Centennial Farm Families manuscript collection in the Society's library also evi-

To the county agricultural agents:

Purdue University is co-operating with the Indiana Historical Society in a joint effort to locate all those farms which have been in the same family for 100 years or more. This means that a direct lineal descendant, man or woman, must still own the century-old farm, or a good portion of it. It is not necessary that the present owner be working the farm, or living on it--just so he owns the farm that his forefather acquired 100 years ago or more.

Please publicize this quest in your county. The information wanted by the Indiana Historical Society is the name and address of present owner, the year the land was entered, and the location of the farm. If you can supply this data, send it to the Secretary, Indiana Historical Society, 408 State Library Building, Indianapolis 4. Or you may ask the farmer concerned to send the information for checking.

The Indiana Historical Society plans to award certificates to these "centennial family farms" next January during Farmers' Week at Purdue University, by way of recognizing the services of these pioneer farm families to the state. Each year thereafter certificates will be awarded as farm familes pass their hundredth anniversary.

The Historical Society will appreciate your assistance in helping gather the necessary information.

Dear Centennial Farm Family:

Here is the Certificate of Recognition presented to you by the Indiana Historical Society honoring your family's possession of your farm for more than 100 years. We congratulate you and assure you of our pleasure in giving this symbol of our interest in your loyalty to the land. If there is any mis-spelling on the Certificate, please return it to us for correction.

The Indiana Historical Society is also offering each of you the opportunity to buy a bronze medallion (shown below) as a souvenir of the occasion. It may be used as a paper weight, hung on the wall, or fastened on the outside door casing. The cost quoted to us is $3.80 apiece, without the upper half of the outside rim. If you wish to have your name cast in the outside rim, there is an additional charge of $1.66. These prices are based on an order of 100.

The Indiana Historical Society makes no profit whatever on these medallions, and you are under no obligation to buy one. We are simply taking orders. Some persons were interested in such a device, and we undertook to find out the cost. If you would like to buy such a medallion for your home, please drop us a card ordering one and indicate whether you want your name on it at the extra cost. SEND NO MONEY. We shall ask you to pay when the medallions are made ready and we have them in hand to mail to you.

Secretary, Indiana Historical Society
408 State Library & Historical Bldg.
Indianapolis 4, Indiana

IHS BASS PHOTO CO. COLLECTION, #230042F

dences Snyder's careful research and attention to detail.

The award for the present-day owners of the Centennial Farms was a certificate, "recognizing the services of these pioneer farm families to the state."[12] In designing the certificate, however, the Society realized that it needed an official seal. The semiannual report of the Society stated that the executive committee was "taking steps to have a seal designed."[13] A Society member and artist, Frederick Polley, submitted designs for the seal, and the executive committee selected one "showing the last Territorial and first State Capitol at Corydon, ringed with the words 'Indiana Historical Society 1830.'"[14] The Certificate of Recognition thus contained the Society's new seal, the owner's name, township, and county, and an acknowledgement of "loyalty to the land."[15] Successful applicants were awarded certificates on 2 January 1948 at a ceremony held during Farmers' Week, part of the annual Agricultural Conference of State Associations at Purdue University. At the time of the ceremony, 1,200 centennial farm families had been identified. According to an article in the *Lafayette Journal and Courier* of 3 January 1948, about one hundred people attended the ceremony.[16] E. L. Butz of Purdue University gave the keynote address, "A Century of Farm Progress," and G. E. Davis gave a reading of James Whitcomb Riley poems. Following remarks by W. O. Lynch, president of the Indiana Historical Society, Peckham awarded the Centennial Farm Family Certificates.

Besides these certificates, the families could purchase bronze medallions, plain or engraved, to "be used as a paper weight, hung on the wall, or fastened on the outside door casting."[17] The medallion had the seal of the Society and the words: "Centennial Farm Family" on it, and the owner's name could be engraved on the outer rim, if desired.

IHS BASS PHOTO CO. COLLECTION, #C684

S. P. Scherer Farm

Opposite: Buckeye wagon, 1934

After the ceremony, "the publicity attending the program at Purdue University aroused many more eligible farm families to make themselves known."[18] The 1947 program proved so successful, in fact, that the Society decided to continue the census another year to recognize all people who had just become aware of the program and those whose land became eligible due to an 1848 land purchase. By the time the program ended in 1951, more than 1,650 farms had been enrolled, and the Society was sending out certificates to the families "as fast as the artist finishes lettering them."[19]

Applications for recognition of a centennial farm asked for fairly dry information: the name of the present owner, the location of the farm, the first family member to own the farm and the date when it was originally acquired, and the relationship of the present owner to the original owner. At the bottom of the form, there were three lines, which the

IHS BASS PHOTO CO. COLLECTION, #284436F2

Barns at Kingans Farm, 1953

applicant could use for "any other information of special interest." This is where the real meaning and value of the award became evident.

Many applicants required more than three lines to write about the family farms they had built and called home for a century or more. Some people provided family genealogies and one applicant sent a copy of his family's history, "When the Bretzes First Trod on the Soil of Southern Indiana." Other applicants mailed in detailed descriptions or newspaper clippings about their farms. One woman sent an original poem, which expressed her feelings about her "house at the end of the lane." Coordinates and tract numbers, addresses and directions were not enough. It was important to the applicants to share facts and to tell stories about their families and their farms:

Some farmers wrote about the value of their farms:

- "The land is in a remarkably high state of fertility."
- "[This is] one of the best producing farms in Indiana."
- "A well on this farm was dug in 1841, is still furnishing the farm with clear, cold water."
- "The original barn built by Washington Long, is still considered one of the best in the county and marks the pioneer farm."

Other people wrote with pride about their ancestors who had first farmed the land:

- "They were good citizens. The four generations were . . . church going people, interested in community welfare and very prosperous, faithful, trustworthy, and energetic people."
- "Dan Wright [first member of the family to own the farm] was the grandfather of Wilbur and Orville Wright—inventors of the airplane."
- "The original home built on this land became an underground railroad station for transporting slaves during the Civil War."
- "The first 4 H calf of Whitley County was born on the Calvin Long farm in Cleveland twp."

Many folks were proud of having all the records for their farms:

- "[We] have original tax receipt for year of 1815."
- "I had the original sheepskin deed until 1917, when a tornado destroyed our house. . . . I found the deed but all lettering is obliterated."
- "Mrs. Terry has still in her possession the original sheepskin grant to great-great-great grandfather Terry signed by Thomas Jefferson."

A few individuals, who wrote particularly long letters, indicated that the award program had awakened a century of memories:

- "You will pardon great length of this letter but it was hard to

IHS CENTENNIAL FARM FAMILIES RECORDS

PROGRAM

of the

Annual

Agricultural Conference

of

STATE ASSOCIATIONS

December 29, 30, & 31, 1947

and January 2, 1948

Purdue University

Lafayette, Indiana

(FRIDAY, January 2—Continued)

10:00-11:30 **Hall of Music.**
Mrs. Charles Krise, Auburn, President, presiding.
Interview: Winners of National 4-H Club Camp Trip, sponsored by state association.
Demonstration: "Be Posture - Wise", Vivian Bales, Hamilton County.
Recognition of Lella R. Gaddis Scholarship winner.
Panel Discussion: National Home Demonstration. Council Meeting.
Business Session.

11:45-12:45 **Memorial Union Building, South Ballroom.**
Buffet Luncheon.
(Luncheon reservations, with money, must be received by Anna P. Rainier, AES Annex, Purdue, on or before December 29).

AFTERNOON

1:00- 3:30 **Hall of Music.**
Mrs. Charles Krise, Auburn, President, presiding.
Music, Home Economics Chorus, DeKalb County.
Installation of 1948 Officers.
Poetry, Esther Kem Thomas.
International Affairs, Dean E. C. Young, Purdue.

INDIANA HISTORICAL SOCIETY
"A Century of Farm Progress"

AFTERNOON

1:30- 4:00 **Eliza Fowler Hall.**
O. G. Lloyd, Purdue, presiding.
Welcome, Dean H. J. Reed, Purdue.
A Century of Farm Progress, E. L. Butz, Purdue.
James Whitcomb Riley Readings, G. E. Davis, Purdue.
Remarks, W. O. Lynch, Bloomington, President, Indiana Historical Society.
Awarding of Centennial Farm Family Certificates, (to families who have owned the same farm for 100 years or more), in charge of H. H. Peckham, Indianapolis, Secretary, Indiana Historical Society.

INDIANA STATE ARCHIVES

KNOX COUNTY HISTORY, 1988

HISTORY OF FAYETTE COUNTY, 1917

Top, left: The Centennial Farm certificates were awarded at the annual Agricultural Conference of State Associations, during the winter of 1947–48. The cover and page 12 of the program are shown here. **Top, right:** The Rodarmel family, 11 October 1906, Knox County. See entry for Rodarmel family on page 92. **Bottom, left:** Thomas J. Timmons' cabin, built in 1828 with two rooms and two fireplaces. A copy of the photograph is in the Williams folder, part of the Timmons family files in the Hoosier Homestead Project Collection at the Indiana State Archives. See entries for Timmons families on page 101. **Bottom, right:** Family reunion at the Handley homestead. See entry for Handley family on page 67.

30312

relate facts as to the more than a century that this farm has been in our name, and in which a great deal of Historical events have transpired and have been handed down from Father to Son & Grand Son."

- "I don't mean to tire you with more information than you have asked for, but there is considerable sentiment, and much history connected with the whole thing."

Some of the memories were rather poignant:

- One applicant told how the old home was her birthplace, and over a hundred years earlier, when the house was new, her mother's birthplace also.
- As she filled in the form, another applicant realized that the day she was applying would have been her mother's 104th birthday.
- One applicant shared the touching story of the original family on her farm: "Peter C. Vannice and Sarah A. Vannice were parents of ten children, rearing them on this farm where they lived 56 years. [They] lived together within a few days of sixty years & in death were not separated he dying Feb 9, 1888, in his 87th year, and she, on Feb 12, in her 79th year. They were laid to rest in the same grave in the old family burying ground west of the house on his farm."

Other memories evoked nostalgia for bygone family members:

- "My father was the oldest of 10 sons and 1 daughter and they have all passed away. I was born December 3rd, 1896, the youngest of our family and the only one living."
- "It is a very dear place to me, and I have tried to make all as nice as possible. I am last of the family living, all born in that house.
- "I am now 85 and owner of most of the old homestead and sole survivor of the family."

IHS BASS PHOTO CO. COLLECTION, #302351

Many letters described the importance of the Centennial Farm award to farm families:

- "My one regret is that my late husband, William McKinnis, could not have shared this honor. Had he lived one month longer, he would have reached the age of ninety-two years. He endured many hardships in his early years of primitive farming, but always maintained his true spirit of honesty and integrity . . . and now I have just received the Centennial Farm Medallion, of which I am very proud and appreciative."
- "One of George's greatest ambitions in life was to own the Benjamin Markey land which he did. And when he saw in

We have the old sheepskin deed made out by Martin Van Buren but signed by his secretary. Can't read his name. The deed has a seal on [it] but can't read it either. I'm copying on the back. It's rather interesting. [Applicant sent a sheet with full transcript of the deed.]

George William Harter
Allen County

Threshing machine, 1912

Opposite: Swine Stockyard, 1912

My great-great grandfather "Thomas Timmons" built the first log cabin in [1831]. . . . The Timmons land grant was the first—the first cabin built and the first cemetery in Benton County, Indiana.

Irwin K. Williams
Benton County

IHS CENTENNIAL FARM FAMILIES RECORDS

"Large pine tree [in photo above] set out in 1870 by Oliver Widney, grandfather of B. V. Widney, county agent at Whitley County."
(Carl D. Carpenter, applicant for Harvester Farms, DeKalb County)

Opposite: S. P. Scherer farm

the paper last spring about the 'Indiana Centennial Family,' he intended to be present at Purdue University in January." [George Markey died 9 May 1947.]

- "I want our farm to be on that list too, for I feel it is a great honor to live on a farm that our grandfather purchased."
- "I am very thankful this is being done for us older ones."
- An applicant who discovered that her family farm was not quite old enough to be a centennial farm, expressed her disappointment and showed what an honor this distinction was to many: "I am sorry to say that my information is not correct. My father came to Indiana in 1852—so I am not the distinguished citizen I thought I was."[20]

When the Centennial Farms program was discontinued in 1951, paperwork concerning the project was placed in the Centennial Farm Families Records, 1947–1950 manuscript collection of the William Henry Smith Memorial Library of the Indiana Historical Society. The applications and other documents were filed in folders by counties. In 1997, while searching for something else, these papers were discovered by Ruth Dorrel, then editor of the Society's quarterly journal, *The Hoosier Genealogist*. Dorrel originally planned to extract the records and publish them in the journal. It became evident, however, that there was too much valuable information for a magazine article; therefore, she proposed this publication.

Dorrel extracted the following information for most of the entries in the "Index to the Original Owners of the Centennial Farms": the name of the original owner(s), the county in which the land was located, the year the land was originally obtained by the family, the name of the person or persons applying for the award, and the relationships of the applicants to the original owner(s). In some cases, paperwork or correspondence was missing for an application. Consequently, some of the data is missing from the entries to Dorrel's index. Applications that may not have been approved for the award are also in the folders. These applications may have been declined because the date of the original purchase of the land was not early enough or because there was not direct lineal descent from the original owner(s) to the owner applying for the award. Dorrel extracted information from these applications as well because the information is valid and useful to genealogists. Their entries are marked with an asterisk (*) in the index.

Information on the application that is not listed in Dorrel's index

IHS BASS PHOTO CO. COLLECTION, #C683

The farm home of Ernest L. Smith & wife Naomi Smith of Honey Creek twp of Vigo county, down on the canal road south of the railroad tower called Springhill, has been in the Smith family since 1816 Dec. 28th & Oct 23, 1835, in separate deeds and abstracts now owned by Ernest L. Smith & wife.

Mrs. Ernest L. Smith
Vigo County

includes the mailing address of the applicant, the legal description of the land, and the names of the persons in the intervening generations between the original owner and the applicant. Researchers may view this information by visiting the Society's library and looking through the Centennial Farm Families Records manuscript collection.

In 1976 the state of Indiana began a Hoosier Homestead Project, and many of the farms listed in the Centennial Farms project have become Hoosier Homestead farms. The Hoosier Homestead applications are housed in the Indiana State Archives. The Archives' Internet database for the Hoosier Homestead collection can be accessed by searching the last names of the descendants of the original owners of the land (those who submitted the applications). Besides the applications, many applicants sent copies of original deeds of land, wills, plots, and other materials to prove one-hundred-year ownership of the land. Thus, the Centennial Farm Families Records collection of the Indiana Historical Society and the Hoosier Homestead collection of the Indiana State Archives are treasure troves for genealogists and family historians researching their Hoosier families and their families' farms.

Notes:

1. "Centennial Farm Families," *Indiana History Bulletin* 24 (December 1947): 265.
2. *Indiana History Bulletin* 24 (January 1947): 6.
3. Ibid. (June 1947): 147.
4. "Centennial Farm Families," *Indiana History Bulletin* 24 (December 1947): 265.
5. *Indiana History Bulletin* 24 (June 1947): 147.
6. Ibid. (May 1947): 132.
7. "Centennial Farm Families," *Indiana History Bulletin* 24 (August 1947): 179.
8. Form letter from Indiana Historical Society and Purdue University to Indiana county agricultural agents, Centennial Farm Families Records, 1947–1950, M0043, Indiana Historical Society, Indianapolis.
9. "Centennial Farm Families," *Indiana History Bulletin* 24 (August 1947): 179.
10. *Indiana History Bulletin* 25 (February 1948): 24.
11. The quotations in this paragraph come from letters in Centennial Farm Families Records.
12. Form letter from Indiana Historical Society and Purdue University to Indiana county agricultural agents, ibid.
13. *Indiana History Bulletin* 24 (June 1947): 148.
14. Ibid. 25 (February 1948): 24.
15. Ibid. (May 1948): 100.
16. *Lafayette Journal and Courier*, 3 January 1948.
17. Centennial Farm Families Records.
18. *Indiana History Bulletin* 25 (May 1948): 100.
19. Ibid.
20. The quotations in this section come from letters in Centennial Farm Families Records. The quotations scattered throughout this volume are from this collection as well. All bracketed material is inserted by the editors.

There's a home that I visit in Memory Land, A house at the end of a lane,
It is very old fashioned, But I love it just the same.

It was my birthplace that grandfather built,
Over one hundred years ago, it was mother's birthplace also. . . .

Father and mother both are gone, we children have much older grown,
Much older than our mother dear, on the day that she left us here.

I still love you dear old home, And you still remain my own,
Perhaps in the land where we never grow old, There will be a replica of you.

Nellie Diehl Higgins
Randolph County

From Tribal and Family Farmers to Part-time and Corporate Farmers: Two Hundred Years of Indiana Agriculture

M. Teresa Baer

The Centennial Farms Award Program of the mid–twentieth century celebrated family farming in Indiana at a time when agricultural economists were beginning to acclaim an exponential growth in farm output. Indeed, advances in mechanization, fertilizers and pesticides, biological development of plants, and agricultural education of farmers spurred more than 100 percent production growth per worker in the American farming industry between 1940 and 1965.[1] The juxtaposition of the two celebrations—that of hailing farms that had been in one family for one hundred years or more and that of hailing farms for a fantastic increase in production—proved ironic. The sad fact was and is that the family farms that created the Midwest could not survive the dramatic lowering of food prices that the increase in production caused. Instead, the acclamations for American agriculture's miraculous progress became the death knell of the family farm, a constant ringing that continues today as the last of America's family farms face extinction.

Irony played a part in the beginnings of midwestern agriculture as well. Pioneers poured across the Appalachian Mountains in hopes of securing enough land to become independent farmers—a dream encouraged by American founding father Thomas Jefferson, who was president at the beginning of the nineteenth century. However, industrialization and the growth of a capitalist economy followed on the heels of the pioneers, dooming the ideal of a nation of small, independent farmers. Nevertheless, the dream looked bright as the Northwest Ordinance of 1787 opened millions of acres of fertile land for purchase in the area that would become the states of Ohio, Indiana, Illinois, Michigan, Wisconsin, and part of Minnesota. Native Americans had lived for generations throughout a goodly portion of this area, hunting in the deep forests, growing crops in the lush bottomlands of numerous streambeds, and conducting international trade through French fur traders. They farmed their lands communally, growing enough produce to supply the families of their tribes and trading partners.[2]

In the early nineteenth century, the Miami Indians, who lived along the upper Wabash, Eel, and Mississinewa Rivers, in what would become north-central Indiana, grew corn, beans, squash, wheat, and some garden vegetables. They gathered medicinal plants such as goldenseal, dog fennel, boneset, and yarrow, as well as edible foods including wild berries and other fruits, greens from mustard, dandelion, and other

It was all woods, some neighbors came in and helped saw down trees to build a log house to live in. He [Robert Dooley] moved his family with a few pieces of furniture when the house was done, a cupboard, he moved from Kentucky is still in the family in good condition.

Bessie F. Nesse
Boone County

plants, and sweeteners such as honey and maple sugar. They hunted numerous animals, too. Although big game like deer, bear, and bison were disappearing from the area due to overhunting, the Miami ate small game such as river otter, muskrat, squirrel, and rabbit, fish of all types, and fowl such as pigeon, duck, and turkey. Some Miami also raised domesticated animals including cattle, pigs, and chickens.[3] In 1765 a visitor to the substantial Miami village of Kekionga, which was in what is now northwestern Ohio, described the village as having forty or fifty cabins that the natives inhabited.[4] By 1817 when Native Americans in the Old Northwest were scattered and weakened by warfare with the United States, a Miami village near Fort Wayne in northern Indiana was described by Indian agent Benjamin F. Stickney as being composed of "rude cabins" and teepees, which were used only during the spring and summer months. The rest of the year the Indians were out in the woods hunting, according to Stickney.[5] Historians today note that the Native Americans' fortunes were declining because they refused to participate in the Jeffersonian ideal of the independent farmer, preferring to remain true to their own ideal—of community—raising, gathering, and hunting food for the group. Had pioneers realized that conglomerates would overtake family farms within two hundred years, they may have decided to coexist with their Native-American neighbors and learn how to retain their way of life while feeding *their* community—the world!

Miami elder Kilsoquah and her son, Anthony Rivarre

IHS INDIANA EXTENSION HOMEMAKERS COLLECTION

As the political fortunes of the Miami and their Native-American brethren declined, American immigrants began arriving into the Indiana Territory. The earliest settlers traveled from the upland South—Kentucky, North Carolina, and Tennessee. Crossing the Ohio River, many of them settled first near Cincinnati before journeying west. In Cincinnati the upland southerners met immigrants from the mid-Atlantic states of Pennsylvania, Virginia, and Maryland. Together these two migrant groups settled the lower two-thirds of Indiana. Following these migrations, people from the New England states, especially Vermont, moved along a northern route into northern and central Indiana.[6] This migration increased after 1830 as the federal government began removing Native Americans from the area. By 1846 only 148 Miami remained on a few large landholdings of between 2,400 acres and ten square miles, where they continued their traditional lifestyle. The Miami population doubled by 1870 due to the return of some western Miami. All other Indian groups were moved beyond the Mississippi River.[7] Meanwhile the pioneer immigration was followed by an enormous influx of foreign-born immigrants at midcentury. The greatest proportion of these came from the German principalities and from Ireland.[8] Thus by 1860, Hoosiers numbered more than 1.35 million and more than 91 percent of them lived in rural areas.[9]

IHS BASS PHOTO CO. COLLECTION, #19188

Plowing with horses, 1910

While most of the early pioneers were white, a tiny portion of them, about 1 percent by 1860, were free people of color—African Americans and people of mixed African, European, and Native-American origins. Many of the black and mixed-race immigrants were free blacks who came up from the South of their own accord, seeking cheap land. Some were brought to the North by their slave owners and set free, while others were fugitive slaves. Similar to their white counterparts, these people of color settled primarily in the countryside, working to establish themselves as independent farmers. Between 1820 and 1850, African Americans created at least thirty settlements in the Old Northwest, most across the central and southern portions of Indiana and Ohio. In 1860, approximately 73 percent of these people were living and working in rural settings.[10]

Pioneers, whether white, black, or of mixed races, led difficult lives, attempting to establish farms, feed and shelter their families, and make enough money to pay the government or the speculators who held titles to their land and to pay taxes on the land. Looking through land deeds and other primary documents, it is not unusual to find that an individual bought a piece of land, lost it, and bought land again within a few months or years. Many early settlers suffered setbacks that forced them to hire out as laborers until they could afford to buy more land or to move farther west where land was still abundant and inexpensive. People of color, who generally had less money and possessions than whites when they arrived in Indiana, suffered particularly from such setbacks.[11]

Many obstacles faced pioneers during the first quarter of the nineteenth century. Land policy stipulated that the federal government sell land in parcels of 640 acres in 1785, a minimum requirement that was reduced to eighty acres by 1820. Thus, most pioneers bought land from speculators at higher prices than government land. Added to this was the high cost of credit in the early Republic when no institutions, regulations, or insurance existed for lending money. Individuals who loaned money could do so for short periods only and at considerable personal risk. Thus credit was scarce, short-term, and expensive. Added to this was the fact that there were few cheap, reliable means of transportation and few markets for farm produce during much of the pioneer period. Even if a farmer could have hung on to a piece of land long enough to produce a surplus for sale, there may have been no way to market the produce.[12]

Pioneers also faced obstacles from nature. Much of Indiana was covered by dense forests, briars, and underbrush, which had to be cleared painstakingly, acre by acre, in order to begin planting crops.

Family legend has it that about the first thing my father [William Stranathan] did was to hew the timbers and build a "woolen mill,"—this not far [from] the junction of Prince William Pike with Wild Cat Creek. . . This Prince William Pike was one of the few, if not the only, state thoroughfare from the south-east part of the new Indiana.

Elizabeth Stranathan Askew
Carroll County

Opposite: Farmers with threshing machine

Shelters for humans and animals were built hastily from the materials at hand and were often crude, drafty, and dark. Insects proliferated inside and out, spreading and exacerbating any disease that swept through an area. With little in the way of medical help and much of it incompetent, disease often spelled death for pioneers.[13]

Faced with overwhelming obstacles and having only crude tools, the earliest Hoosier farmers, like their native neighbors, practiced subsistence farming. The average pioneer came into Indiana with less than $200, one horse, a cow, a few pigs, cornmeal or wheat flour, salt, cooking implements, blankets, clothing, a rifle, and scant tools, including an ax, a hoe, and a scythe. Wealthy pioneers might also have a wagon and a team of horses or oxen. Therefore, the average pioneer family could cultivate only ten to twenty acres in a decade or two, growing corn, squash, potatoes, wheat, and vegetables. Hunting, fishing, and gathering supplemented the pioneer family's diet, while the sale of lumber, and oftentimes the sale of improved land, helped to supplement cash incomes.[14]

Advances in farm tools from the 1820s through the 1860s as well as improvements in transportation encouraged early Hoosier farmers to move from subsistence to commercial farming. The most important tool was the plow, which broke and turned the soil. American pioneers had a difficult time breaking through thickly matted roots, more than a foot deep in the soil. Lightweight plows could not handle newly cleared fields, so farmers who could, fell back on heavy plows with iron shares for cutting into the soil and wooden moldboards for turning the earth. Such plows required tremendous animal power to use, however. Two Illinois blacksmiths, John Lane and John Deere, improved upon the plow design in the 1830s. Deere's design, "a one-piece wrought iron plow with a cutting edge of steel on the share," became so popular that Deere and a partner began mass producing about 1,000 plows per year. Mass production lowered the cost so that Deere's plow became increasingly affordable, greatly enhancing the production capabilities of small midwestern farmers.[15]

Other tools that early Hoosier farmers adopted included the harrow, cultivator, cradle, reaper, and thresher. Harrows had been around since before Roman days in the form of tree branches dragged across broken soil to loosen it up for planting. By the 1840s, midwestern farmers were using hinged harrows, with metal spikes that could be pulled by two horses. Cultivators were mounted on wheels in order to control the depth of the tines or small shovel plows, and they could break up earth or cover seeds. The cradle replaced the scythe, cutting twice as much grain as a scythe and bearing an attachment that caught the grain. By the 1850s the cradle was being replaced by the McCormick Reaper, a machine with shears that cut grain with a vibrating motion and collected it on a platform. Cyrus McCormick of Virginia patented his original design in 1834, made continuous improvements until 1850, and was selling 40,000 reapers annually by 1860. After reaping grain, threshers separated the grain from the chaff. A stationary model, designed by Hiram and John Pitt in 1836 proved to be inexpensive and was widely used by farmers in the Midwest within the next decade.[16]

By 1860 more than 8.25 million acres of land in Indiana was under cultivation. Fewer than 132,000 farms produced 69 million bushels of corn, 15 million bushels of wheat, more than 7 million pounds of tobacco, 5 million bushels of oats, more than 3.8 million bushels of potatoes, 400,000 bushels of rye, and more than $1 million worth of orchard

IHS BASS PHOTO CO. COLLECTION, #20788

Sheaves of corn

products. On average, each Hoosier farm family owned 19 hogs, 16 sheep, several chickens, 4 stock cattle, 4 dairy cows, 3 horses, 1 ox, and perhaps a mule. The sheep supplied wool, which the farm women and children spun and then wove into cloth. The chickens and dairy cows were essential for eggs and for the milk products that Hoosier farm women made for their families' use and sold to neighbors and at the local market. Commercially, on the eve of the Civil War Indiana had become one of the top two wheat producers and one of the top four corn producers in the country. Farmers fed much of the corn surplus to livestock, a natural consequence of the fact that Hoosiers owned more hogs than any other state at this time.[17]

Hoosier farmers could not have begun to farm commercially without improvements in transportation. Prior to 1820 few means of transport existed. Cattle and hogs were driven east, particularly to Madison, Indiana, and Cincinnati, Ohio, for sale and slaughter. Corn was ground at the many gristmills that had sprouted in Indiana, cured into whiskey, and carried by packhorse to towns. Grain and livestock were also floated down the Ohio and Mississippi Rivers. The advent of steamboats in the 1820s greatly facilitated the latter transportation method. Steamboats in the Midwest were built to carry heavy cargo and could run in water as shallow as three feet. The opening of the Erie Canal also made it possible to move agricultural produce from the Midwest to New York City, expanding the farm markets to New England and Europe. The state of Indiana tried and failed to create a system of canals during the 1830s, but Hoosiers did build 700 miles of plank roads by 1860. Plank roads, however, proved dangerous because they were slippery and the planks tended to overturn. The single most important transportation improvement was the railroad. In 1850 there were 228 miles of track in Indiana; by 1860, 2,100 miles of track ran through the Hoosier State, connecting it to St. Louis, Chicago, and the East Coast.[18]

While the improvements in farm tools and transportation fueled the growth of commercial farming by allowing farmers to produce considerable surpluses, they also fueled the beginning of the end of the Jeffersonian dream. Instead of a nation of small, independent, self-sufficient farmers, each with his or her private domain, the United States began the process of aggregation. During the second half of the nineteenth century, farming became an industry led by market forces that encouraged each area of the country to produce what was best suited to grow in the region (mostly corn, hogs, and cattle in Indiana) and to "trade" the produce over vast transportation networks with large urban centers and middlemen acting as conduits. An expanded railway system that sported refrigerated cars facilitated the farm sector as did the advent of the canning industry to preserve perishable food items. Farmers who could not compete successfully in this economic landscape were forced out of the market or became renters, working on the land of wealthier, independent farmers. At the other end of the spectrum, the returns to production of improved equipment tapered off after the Civil War and successful farmers were forced to place increasing amounts of land under cultivation in order to realize a profit.

Farmers and farm laborers, including the sons of midwestern farmers who did not inherit family land, settled the West during the latter half of the nineteenth century, bringing millions more acres of land into the American agricultural industry and producing more and more food. The result was overproduction. As farmers bought enhanced farm equipment that required animal rather than human power, paid increasing taxes on increasing amounts of improved land, and paid substantial transportation and handling charges to move their products across the continent, the revenue from their products declined. In the antebellum period, a huge influx of immigrants into the United States provided increasing demand for agricultural products. But immigration declined sharply during the Civil War and it did not increase again dramatically until the end of the nineteenth century. In the meantime, farmers grew more food but received less money for it.[19]

Besides lower profits, there were several consequences to the overproduction of agricultural products and the relatively higher cost of production and taxes. As mentioned above, some people became renters or farm laborers while others were pushed out of the farming industry altogether. Thus, in 1900 more than one-third of Hoosiers lived and worked in cities. Black farmers had a more difficult time than others since they had started their farms with less money and possessions, had been able to buy less land, and had subsequently been unable to compete as successfully as most white farmers. Blacks tended to turn to the cities for their livelihoods sooner than whites. Coupling this fact with the influx of ex-slaves into the North after the Civil War, meant that little more than a third of all blacks in the states of the Old Northwest were farmers by 1900. Few of Indiana's approximately 440 Native Americans were farming by the start of the twentieth century either. Federal legislation in 1873 and 1887 had divided the Miami landholdings into individual farms and sold off the "surplus land," leaving Indiana's natives without the resources necessary to lead their customary, subsistence way of life. With little cash to invest, little understanding of commercial farming, and little help from surrounding neighbors and local governments who tended to take advantage of their ignorance, most of the Miami land was lost by 1900.[20]

IHS INDIANA EXTENSION HOMEMAKERS COLLECTION

Our grandparents [Mr. and Mrs. Nathaniel Hardester Modesitt] came from West Virginia to Indiana by boat—down the Ohio River and up the Wabash to Terre Haute, then overland by ox-cart.

Annie M. Modesitt
and Ruth L. Modesitt
Clay County

> **My Grand Father settled on this Farm in the year 1808. We also have a grand Father's clock built in the House we live in, in the year 1816[; it] is 8½ feet tall.**
>
> Ira C. Fouts
> *Clark County*

Many of those who continued farming throughout the end of the nineteenth century formed political and business groups in order to alleviate the challenges that beset them. The Patrons of Husbandry, better known as the Grange, and farmers' alliances formed political parties and installed representatives in state and federal governments. Two of the goals of these groups were to regulate the railroads and to establish cooperative farming ventures that would cut out the middlemen and ensure higher prices for farmers' products. These goals were not achieved. The railroad regulatory laws were overturned by the courts and most nineteenth-century cooperatives were short-lived and financially unsuccessful. However, the right of states to regulate the railroads was established and farmers learned valuable lessons through cooperative farming arrangements that would benefit subsequent generations of farmers.[21]

One of the lessons that the government and farmers learned during the latter half of the nineteenth century was that in order for farms to remain viable it was essential to learn more about agriculture—its inputs such as seeds and soil and its processes including the nurturing of plants and the breeding of animals. In 1862 Congress passed and Abraham Lincoln signed into law a bill establishing the U.S. Department of Agriculture (USDA). The USDA's mandate comprised research on plants, soils, and manure (fertilizers), answering farmers' inquiries, and collecting and disseminating agricultural data through publications, a professorship, a library, and a museum. In 1887 Congress also passed the Hatch Act, whereby the federal government provided financial support for scientific research on experimental farms throughout the country, support that was used by Purdue University in Indiana.[22]

Farmers turned to the allianced groups in their states for education. In Indiana the Grange and other agrarian organizations distributed information about the latest farming implements and practices and encouraged farmers to keep accounts and determine areas of profit and loss within their operations. In addition, Purdue University (named for its original patron, Lafayette businessman John Purdue) began conducting agricultural research and providing education via farmers' institutes and other extension services in the 1880s. Some of the earliest experiments at Purdue concerned the use of manure and other fertilizers. By the 1870s, Hoosier farmers were realizing that their traditional strategy of growing a particular crop such as corn in a field, year after year, depleted the soil's productivity. They used manure to enrich the soil, but the process was too labor intensive to be effective until a manure spreader was invented and widely used in the 1880s. During this period, some southern Indiana farmers also began using commercial fertilizers. Experiments by Professor W. C. Latta at Purdue initially indicated that the use of commercial fertilizers on soil that had been badly depleted was not economically feasible because the price of fertilizer far outweighed the gain in produce. However, after continuing its experiments for a four-year period, Purdue began encouraging the use of fertilizers for corn and wheat. Thus, after 1891 Indiana farmers widely adopted the use of fertilizers.[23]

The farming sector of the American economy was set to burgeon at the dawn of the twentieth century. Consolidation of farmland into large farms and farm corporations would play a key role in agriculture's success during the century and many factors that spurred consolidation would be key factors in the success of the industry, as well. But success for the industry would be disastrous for thousands of family farmers, a situation that would worsen as the century ensued and create deep divisions between corporate and family farmers by century's end. At the beginning of the 1900s, however, farmers, agricultural schools, farm cooperatives, and some politicians looked toward developing a sustainable agricultural base—whether on an individual farm, throughout a state, or nationwide. Education was an important factor in American agriculture's success. Accordingly, Purdue's Farm Institutes became a strong tradition throughout Indiana. Added to this were agricultural exhibition trains, which ran until 1947, and the beginning of Purdue's Agricultural Extension Department, which made its debut in 1910–11. Purdue initiated regional and statewide conferences, research and development efforts on seed, produce, dairy products, and equipment, children's agricultural clubs, and competitions regarding farm products. Purdue's leadership was part of an agricultural trend in the United States. Private and government scientists were beginning to learn about the chemical compositions of different soils and plants, about breeding hardier plants and animals, and about plant and animal diseases—following the pioneering efforts of Justus Liebig, a German who wrote about the effects of chemistry in agriculture; Eben N. Horsford, John A. Porter, and Samuel W. Johnson, Americans who studied with Liebig; and American John P. Norton, who studied in Edinburgh. Through agricultural colleges such as Purdue, American agricultural scientists were sharing their findings with farmers throughout the country.[24]

Scientific and technological breakthroughs helped American farmers increase production by a modest 8 percent over the course of the first decade of the twentieth century. Scientific advances included nitrogen and lime fertilizers, advances in plant breeding, and control of animal diseases, such as hog cholera, for which an effective vaccine was invented in 1903. Farmers also benefited from larger and more efficient seeding machines, cultivators, and harvesters. Overall production increased at 9 percent from 1910 to 1920, because the last of America's arable land was put to use for agriculture. However, the per-acre productivity of America's farms actually declined during the second decade of the century, because, once again, the benefits of the new technologies had reached their limits.[25]

IHS DAVID PEAT COLLECTION

Advance Rumely Separator

The single most important reason for the strengthening of American agriculture prior to World War I was the increase in demand. Demand rose in the United States due to renewed waves of immigration, much of it from eastern and southern Europe.[26] Indeed, between 1900 and 1910, the U.S. population increased two-and-a-half times as much as farm production. Demand also rose dramatically in Europe during World War I. Thus, farm prices climbed upward steadily from 1897 to

Burr Braden paid passage to Republic of Liberia, Africa, of his 100 slaves, from London Co. Virginia, before settling in Clinton County, Ind.

Wm. R. Braden
Clinton County

1910, held steady at a high level through 1914, and more than doubled during the war years.[27]

Indiana's farmers participated in American efforts for the First World War, plowing up pastureland to help meet increased demand for grain at home and abroad. Ten thousand new silos were built between 1917 and 1918 to hold the grain. Hoosiers in all walks of life grew victory gardens and canned their produce to help the war effort. At the same time Hoosier farmers enjoyed soaring prices for their products and sought to capitalize on the profits by buying more equipment and land, even though the price of land was grossly inflated.[28]

Between 1915 and the early part of 1920 the price of land in the United States was at a premium for several reasons: the soaring profits gained in the farming industry from 1897 to 1920, the scarcity of farmland in the United States, where nearly all arable land was under cultivation, and the fact that the increases in production from improved technology were again stagnant. All farmers were eager to make the most of the windfall in the industry and so they were eager to buy more land—at this time the only unit of production that could increase productivity. Of course what they did not know—could not foresee—was that the windfall would soon come to an abrupt end. Unaware of the danger, many farmers bought as much land as they could on credit, expecting future profits to more than pay off their debts.[29]

The end of the war signaled the end of large exports overseas, however. Immigration declined considerably after 1920 and was negligible by the end of the decade due to governmental restrictions. Thus farm prices plummeted to about half of their initial worth during 1920 and recovered only partially during the ensuing decade. At the same time, the value of nonfarm products rose, forcing farmers to pay off credit debts with declining profits. All farmers suffered at this time and many farmers failed altogether. By 1920, due to the influx of immigrants who settled mostly in the cities as well as the sharp changes in the profitability of agriculture, more than half of all Americans, including Hoosiers, lived in urban areas. At this time, the number of blacks and other people of color who lived and worked in the countryside of the Old Northwest had dwindled to 14 percent. In northern Indiana, the Miami owned less than 1,000 acres and nearly all of the tiny population of Native Americans lived in cities.[30]

The years between the First and Second World Wars were heartbreaking for Indiana farmers, just as they were for most farmers across the country. Following slight gains in the prices of grain, livestock, and food products during the mid-1920s, prices declined again by 1928. Land prices were low as well. The stock market crash in 1929 dealt the final blow to many farmers, including Hoosiers who lost their land, their homes, their equipment, and their way of life forever. By 1932 Indiana's farm-cash incomes were about one-third of what they had been in the desperate year of 1929. By January 1933 delinquent property taxes in the state totaled more than $25 million, with only nineteen counties showing a less than 10 percent delinquency rate. Americans were in the throes of the Great Depression, a period when numerous mortgage foreclosures led to violence and even suicide across the nation.[31]

As agricultural bust years settled in, Hoosier farmers sought relief from low incomes, inadequate credit arrangements, and taxes that were high relative to the declining value of their land. The Indiana Farmers Federation organized in Indianapolis in March 1919, with representatives

IHS INDIANA EXTENSION HOMEMAKERS COLLECTION

from around the state, and became part of a national federation in November. In 1923 Indiana's federation became the Indiana Farm Bureau with the slogan, "Equality for Agriculture," a political philosophy shared by farmers and farm organizations across America to the present day. In alliance with several other groups, including the Indiana Grange, the Farm Bureau worked as a cooperative, purchasing members' farm surpluses in order to improve the members' bargaining position on pricing. This activity was part of a national movement and a resurgence of tactics that had failed a half century before. Nevertheless the movement was effecting federal legislative action designed to establish prices for the different food portions of the agricultural industry that would afford all farmers a rate of pay equitable to other sectors of the American economy, such as manufacturing. Initially the movement failed. In 1924, 1925, and 1926 the legislation was defeated because it was too complicated, was unconstitutional, and was aimed at price fixing. When the bill finally passed with changes in 1927 and 1928, it was vetoed by President Calvin Coolidge. Coolidge's successor, Herbert Hoover, was opposed to this farm legislation as well, and the bill was tabled following his election.[32]

In Indiana, on the other hand, the Farm Bureau had success as a cooperative that purchased farm supplies for Hoosier farmers at lowered costs. The bureau bought fertilizer, petroleum products, and feed from 1927 to 1930 for its members. Beginning in 1930 the bureau supplied lubricants for farmers from its oil-blending plant, fertilizer from its fertilizer plants, feed from its mills, paint from its factory, and insurance from its two companies. The bureau also participated in the development of Indiana communities and monitored political activities that dealt with agriculture.[33]

Since the midwestern pioneer days, American farmers had suffered from a lack of financial credit. The federal government finally began addressing this problem in 1908 when the Country Life Commission, appointed by President Theodore Roosevelt, called for adequate and fair loans for farmers. The idea became a major issue in the presidential campaigns of 1912, and President Woodrow Wilson and the Congress of 1913 sent a commission to Europe to study agricultural credit. American agricultural groups argued with the commission's findings, but finally, in 1916 Congress and President Wilson enacted a Farm Loan Act. The bill "authorized both a cooperative system of twelve federal land banks and a system of joint-stock land banks to be organized with private capital for profit," with the federal government providing part of the start-up capital and regulations for the interest rates the banks could charge. The private joint-stock land banks fell victim to the worsening economic conditions of the 1920s and early 1930s and were liquidated by 1933. However, because of the desperate plight of farmers, Congress passed the Emergency Farm Mortgage Act in 1933, which enabled the federal land banks to make short- and intermediate-term loans to farmers and to farm cooperatives. The federal land banks also began increasing the number of loans to farmers and extended "long-term amortized loans secured by first mortgages on farms." The federal government helped to fund these loans, mostly by the sale of land bank bonds.[34]

At the same time that the federal government was writing legislation to assist farmers with credit problems, Indiana farmers were calling for a suspension of mortgage foreclosures and property taxes. The Indiana Farm Bureau gathered 10,000 farmers together in Indianapolis in February 1933, where they marched on the statehouse, presenting Gov. Paul V. McNutt with petitions bearing 50,000 signatures. The petitions requested that the state government enact legislation to create income, sales, and intangible taxes in order to prop up the value of real property. After being told that the governor was "sympathetic to their objectives," the farmers went home. As a result, the Indiana General Assembly quickly passed a measure initiated by McNutt, a gross income tax, which essentially combined sales and income taxes. This act enabled Indiana to obtain revenue at pre–Great Depression levels, while reducing the tax burden of property/landholders.[35]

The worsening economic depression finally moved the federal government to help agriculture achieve parity with the other sectors of the economy. One of the first actions of Franklin D. Roosevelt and the Congress of 1933 was to pass the Agricultural Adjustment Act (AAA), which "authorized the federal government to enter into agreements with farmers to control production by reducing acreages devoted to basic crops [corn, wheat, and cotton], to store crops on the farm and make payment advances on them, to enter into marketing agreements with producers and handlers in order to stabilize product prices, and to levy processing taxes as a means of financing the crop reduction program."[36] The AAA was deemed unconstitutional by the Supreme Court in 1936 due to its taxing provision. Thereafter the AAA of 1938 stipulated that farmers would be paid to hold down the production of corn, wheat, and cotton as well as tobacco and rice and to sell their products to the government rather than handlers, while the government became more involved in the regulation of both domestic and foreign agricultural goods. The 1938 act became "part of permanent agricultural legislation," which means that during any particular period, unless Congress enacts a temporary or short-term law that supersedes all or part of its provisions, the permanent provisions of the 1938 act are in effect.[37]

The Agricultural Adjustment acts of the 1930s revitalized Indiana agriculture. Purdue's Cooperative Extension Service held daily sessions to explain the production control programs to farmers. As the number of mortgage foreclosures and lawsuits to collect farm debts declined, the confidence of Hoosier farmers was restored. In 1935 alone Hoosier farmers received about $19.4 million from the AAA program and farm prices rose considerably—livestock and corn prices nearly doubled within one year and wheat prices moved up slightly. Another of Roosevelt's programs that helped Indiana farmers was the Rural Electrification Administration or REA, which became a permanent agency in the USDA in 1944. The REA "makes low-interest, self-liquidating loans to farmers, farm cooperatives, and utility districts to finance rural electrifi-

Farm name Tlinkatonka[; there was an] Indian camp on [the] farm. [This] could be the home of Tlinkatonka, a descendent of the [----] wife of Kill Buck an indian who saved a white family from the Pigeon Roost massacre.

Roscoe Bright
Decatur County

> **My grandfather [Thomas Horrall] fought in the battle of Tippecanoe and for his services was given the chance to purchase this ground.**
>
> Edith Horrall Bingham
> *Daviess County*

cation projects and improve rural telephone service." Acting on the REA, the Indiana General Assembly created cooperatives for the distribution of electricity through the Indiana REMC Act of 1935. Before this legislation was enacted, only 11 to 12 percent of Indiana's farms had electricity. By 1965, nearly all Hoosier farms had electricity.[38]

It is arguable whether Roosevelt's programs or World War II pulled American agriculture out of the Great Depression. After the war began in Europe in 1939, the demand, and therefore the prices, for U.S. farm goods skyrocketed. Due to this positive economic environment, the AAA's price supports went unused from 1940 to 1948 and from 1950 to 1951 for the five "basic" commodities and peanuts. However, the act was extended to cover fourteen other farm products, including milk and butterfat, chickens and eggs, turkeys and hogs, certain dry peas and beans, soybeans, flaxseed, and peanuts used for oil, American Egyptian cotton, and potatoes and sweet potatoes.[39]

Whether or not the AAA helped end the depression, it certainly had huge effects on the agricultural sector of the post–World War II American economy. The act's premise that the government has the right to support farm prices and income has been at the foundation of all subsequent agricultural legislation. At the same time, searing debates have transpired from time to time over this premise. Willard W. Cochrane, a Harvard-educated economist (whose career included work in the Minnesota state government, the USDA and other federal positions, the United Nations, the Ford Foundation, and professorships at Pennsylvania State University and the University of Minnesota), described the history of the debate from 1947 to the late 1970s in his text, *The Development of American Agriculture: A Historical Analysis*. According to Cochrane, there were two major sides to the debate during the mid–twentieth century: those for little or no government price and income supports, whose advocates included Republican party leaders, agribusinessmen, and most economists; and those who wanted to retain or increase government price and income supports, including Democrats from the South and the Plains states, farm organization leaders, and some union leaders and government economists. Although the main provisions of the 1938 act continued, between 1947 and 1977 the amount of support changed from year to year and from Congress to Congress. Cochrane argued that the governmental income and price supports together with the widespread use of twentieth-century technological advances hastened the consolidation of America's farmland.[40]

Certainly new technology had huge effects on agriculture after World War II. Purdue University's report, *One Hundred and Fifty Years of Indiana Agriculture*, which covers Hoosier farm history from 1816 to 1966, describes the technological innovations and scientific discoveries that accelerated the pace of growth in Indiana's farm industry after 1945: "mechanization, improved crop varieties, better livestock, accelerated use of fertilizer [nitrogen, phosphate, potash] and lime, new agricultural chemicals [including pesticides and weed repellants] . . . plus larger farms, [and the] substitution of capital for labor." While the federal farm acts allowed many farmers the opportunity to invest in new technologies, the pace of agriculturally related scientific advancement during the mid–twentieth century allowed farmers who purchased the technologies to increase their yields per acre exponentially. For example, Hoosier corn growers harvested about 39 bushels per acre in 1915 and an astounding

94 bushels per acre in 1964. In the same years, wheat growers produced 18 as opposed to 34 bushels per acre. Between 1930 and 1965, hog production, milk production per cow, and egg production per hen more than doubled. All in all, from 1915 to 1965, the output per American farm worker nearly quadrupled. Stated another way, in 1930, an average American farm fed 10 people; in 1965, the average American farm fed 33 people and the average corporate farm fed between 100 to 200 people. Indeed, by 1966 less than 6 percent of the American population worked on farms, feeding most of the American population of approximately 180 million people and many other people around the globe![41]

The new technologies that fueled this miraculous growth in farm production began with motorized tractors. Hoosier farmers owned approximately 2,500 tractors in 1917 along with hundreds of thousands of horses and mules. By 1964 more than 181,000 tractors had replaced animal power on Indiana farms, pulling various types of farm machinery that were operated with hydraulic controls, including cultivators, planters, and mowers. Shredders and cutters helped farmers improve brushlands and pulverize cornstalks. Spraying equipment helped farmers to fertilize and debug crops. On the largest farms Hoosiers were also using motorized harvest machines such as combines for cereal grains and picker-shellers for corn by the 1960s. With the use of electricity farm women's chores were also made more efficient and productive. Electricity fueled washers, dryers, cooking stoves, furnaces, sewing machines, freezers, and many other modern household conveniences, as well as enabling farm families to light their homes and barns and to pump water, operate milking machines, run mechanical feeders, and so forth.[42]

The types of plants that Indiana grew changed over the course of the twentieth century. Hoosier farmers planted open-pollinated corn varieties early on, but with Purdue's encouragement, switched to hybrids that were stronger, higher yielding, and disease resistant. In addition, high lysine corn, created by Purdue scientists in the 1960s, proved to be a better source of protein than earlier corn strains. During the 1920s, Hoosiers also began growing soybeans, a Southeast Asian plant, at Purdue's suggestion. A bulletin from the university's agricultural experiment station in March 1920 stated that soybeans were good for the soil, more nutritious and higher yielding than any grain or forage material for animals except alfalfa, and that they would be valuable as a balance to corn in the diet of livestock. Later in the century, soybeans were also used for human consumption in vegetable oil, as a protein substitute, and for industrial purposes such as plastic and paper products. By 1965 nearly three million acres in Indiana were planted in soybeans for a record crop of 83 million bushels. While corn and soybeans were the

IHS INDIANA EXTENSION HOMEMAKERS COLLECTION

On this farm in 1839 Mr. Smith and Mr. [Reuben Grant] Daniels erected the first schoolhouse in the township. Miss Murray & Miss Phelps were the first two teachers. The first wheat in Smithfield Township was sowed on this farm.

Hazel Daniels Wittmer
DeKalb County

IHS INDIANA EXTENSION HOMEMAKERS COLLECTION

Hoosier State's two largest crops in the 1960s, diversification was evident in the state's production of wheat, tobacco, strawberries, melons, and popcorn.[43]

The types of animals that Hoosier farmers raised changed over the course of the twentieth century just as the types of plants changed. Hogs were bred for lard as well as meat before World War I, but after this, they were increasingly grown exclusively for meat purposes. They were also bred to be bigger and bigger until it was determined that the largest hogs, weighing between 500 and 1,000 pounds, had low meat quality. Subsequently hogs were bred and raised to produce lean meat. An important strategy that helped farmers raise hogs with high meat quality was the feed the animals ate. As described above, corn and other grains became increasingly more healthful over the years, helping to produce leaner, larger hogs. Hog production amounted to about one-fourth of all cash farm income in Indiana in 1964. Cattle were also important to Hoosier agriculture during the mid–twentieth century. Although the number of dairy cattle in Indiana declined from 1950 to 1960, Hoosier cows produced more milk and butterfat than the national average in 1963 and 1964. Hoosiers also started raising more beef cattle, especially in the hilly lands of southern Indiana where crop farming was less productive than in the central and northern plains portions of the state.[44]

Purdue's report cites larger farms and the substitution of capital for labor as two major reasons why Hoosier agriculture was such a success story in the mid–twentieth century. Cochrane states that government income and price supports and the twentieth century's increasingly advanced technologies—including machinery, improved crops and livestock, and agricultural chemicals—induced farmers to invest in more land and more of the new technologies in order to realize a profit. He describes a vicious spiral whereby the wealthier and more ambitious farmers use new technologies quickly, realizing a profit for a short while until the new technologies become widely used. Those farmers who cannot afford or do not choose to use the new technologies are marginalized until debt forces them to sell. The other farmers buy up this land in order to gain profits by having more land either under cultivation or in use for livestock. The buying, consolidation, and continuing improvement of arable land—a finite source that was in total usage in America as of the 1920s—drives up the price of land. Eventually the cost of land outweighs the profits that are taken in. At the same time, increased aggregate production drives down agricultural prices, and the farmers who are still in business—less farmers farming nearly the same amount of land as before—no longer realize a profit.

Government income and price supports allow enterprising, rela-

Table 1: An Overview of Indiana Farms, 1940 to 1997

	1940	1959	1978	1997
Indiana land in farms *(in acres)*	19,800,778	18,613,046	16,824,438	15,111,022
Average size of Indiana farms *(in acres)*	111	148	208	261
Number of Indiana farms	184,549	128,160	82,483	57,916

Table 2: Farm Size Versus Number of Farms, 1940 to 1997

Farm Size	1940	1959	1978	1997
1–49 acres	56,420	33,559	21,772	18,170
50–179 acres	98,422	39,652	31,559	19,913
180–499 acres	28,215	34,047	20,984	11,099
500–999 acres	1,332	2,906	6,552	5,268
1,000–1,999 acres	160	256	1,452	2,753

Table 3: Number of Farms of 2,000 Acres or More, 1964 to 1997

Farm Size	1964	1982	1997
2,000+ acres	60	251	713

Statistics for these tables were derived from statistical reports by Ralph Gann, state statistician, and Steve Wilson, deputy state statistician, Indiana Agricultural Statistics Service, National Agricultural Statistics Service. The reports are available at the following web site as of June 2002: www.nass.usda.gov/in/.

tively wealthy farmers to purchase more equipment, chemicals, seed, animals, and land, which simply accelerates the process of balancing the cost of inputs with the returns to investment. Thus, farmers increasingly try to purchase more land and to increase the yields on the crops and the animals they raise in order to make money. While all farmers use the same strategy, the largest, wealthiest farmers continue to push the smaller farmers out of business. The result is increasingly larger farms with increasingly fewer farmers. In Indiana at least, the result is also a growing number of farmers who work outside the farm to produce income, becoming in effect, part-time farmers. Purdue reported that more than 1/3 of Hoosier farmers worked off their farms one hundred days or more by the 1960s. At the same time, the production capabilities of the aggregate agricultural industry grow so long as technology continues to improve and to be used by most farmers. This last point is verified by Indiana's miraculous growth in yield per acre and yield per animal during the mid–twentieth century.[45]

The fact that farms in Indiana became increasingly larger and that fewer people were farming over the latter part of the twentieth century can be seen in the tables opposite. Table 1 depicts an overview of Indiana farms, showing the amount of land used for farming, the average size of an Indiana farm, and the number of farms in Indiana in roughly twenty-year intervals between 1940 and 1997. The amount of land fell by more than 4.5 million acres in this fifty-seven-year period. Purdue's agricultural history accounted for the loss of farmland through 1966, citing the construction of roadways as well as shopping, industrial, and residential complexes. Without scientific evidence, anyone who has lived in Indiana in the last thirty-five years has witnessed a continuation of the building trend, which may explain the cumulative loss of Indiana farmland in the latter part of the twentieth century.[46]

Table 1 also shows the average size of Indiana farms, which contained 2 1/3 times more acreage in 1997 than in 1940. Finally, the first table shows a drastic decline in the number of Hoosier farms—more than 2/3 of Indiana's farms disappeared between 1940 and 1997. This table shows the trend clearly: From 1940 to 1997, Hoosier farms were getting larger while fewer Hoosiers were owning farms.

Milton Thornburgh, a grandson of Henry, owned this farm when he was the first owner of a horseless carriage in Fayette county—a Locomobile purchased in August 5, 1900. This farm was named Mt. Eschol by Henry Thornburgh after the biblical mountain mention[ed] in the thirteenth chapter of numbers.

Carl Caldwell
Fayette County

IHS BASS PHOTO CO. COLLECTION, #C687

Springdale Farms

Opposite: Sow and piglets, McMillan Feed Mills

Tables 2 and 3 show changes in the numbers of Indiana farms possessing particular amounts of acreage: Table 2 shows farms between 1 and 1,999 acres over the period 1940 to 1997, and Table 3 shows farms of 2,000 acres or more from 1964 to 1997. Interesting in these statistics is the drastically declining number of farms of 1 to 999 acres during this period when the number of farms of 1,000 acres or more rises steadily. Looking more closely, a subtler pattern within this overall scheme emerges: the smaller the farms the faster the drop in the number of farms, while the reverse is true for the larger farms. Thus, the number of farms of less than 180 acres decreases rapidly over the fifty-seven years. The number of midsize farms of 180 to 499 acres rises in the first nineteen-year period but falls steadily afterwards; likewise, the number of farms of between 500 to 999 acres increases the first forty years, then starts dropping slowly afterwards. Finally, the numbers of the largest farms increase over all intervals. There were 17 times more farms of 1,000 to 1,999 acres in 1997 than in 1940 and there were nearly 12 times more farms with 2,000+ acres in 1997 than in 1964. These statistics tell the story of farm consolidation in Indiana over the last sixty years of the twentieth century: small, independent farms were disappearing in the wake of large farm conglomerates.

In the last quarter of the twentieth century, Jefferson's dream was rapidly dying, plagued by the growing shadow of American agribusiness. A period of hastening consolidation occurred during the 1970s, an outgrowth of another boom and bust cycle. As in the past, a boom in demand for U.S. food products raised farmers' revenue and inflated the price of land as farmers bought as many acres as possible for cultivating. Foreign demand, which had been important in the two world wars, was the driving force again, but for different reasons. In 1971 President Richard M. Nixon devalued the dollar, making American products more attractive to foreigners. This was followed quickly by détente between America and the Soviet Union, which allowed the latter country to start purchasing American wheat and feed grains. During the 1970s the developing countries finally became wealthy enough to afford American food, and western European countries bought more U.S. agricultural products because adverse weather had lessened their food supplies. In addition, the 1970s was a period of high overall inflation, which pushed the cost of American agricultural products upwards. "Indeed, between 1972 and 1974 wheat prices doubled and corn prices tripled."[47]

Just as in the past, farmers sought to increase production in order to increase profits. The federal government encouraged this strategy by lifting production quotas. Some agricultural lending institutions gave credit only to large agribusinesses or to those farmers who planned

My mother lived hear [*sic*] til she died at age 92 years. I myself have lived here 77 ¼ years. Sleeping in the same room that I was Borned in.

Lizzie McPheeters
Floyd County

bold production increases. Many lenders loaned money against inflated land prices, setting farmers up for deep debt situations if the boom ended. As in the past, many farmers risked their farms in these investment ventures in order to squeeze gains out of what had traditionally been a tight market.[48]

Once again, numerous farmers lost the gamble. In 1977 farm prices began falling rapidly. A decline in foreign demand initiated a bust period in agriculture. This time, however, American farmers were faced with the effects of changes in the economies of foreign countries that made these nations competitors with rather than dependents of American agricultural products. When foreign demand had risen dramatically in the early 1970s, farmers around the world had answered the opportunity. By 1977 climatic problems had ended in western Europe and the effects of a stimulated global agricultural economy were being felt in the usual forms—overproduction and falling prices. Political decisions worsened the situation. In 1979 the Federal Reserve Board raised interest rates in order to curb inflation. Two years later President Ronald Reagan enacted a massive tax cut that pushed interest rates up again. Higher interest rates raised the cost of farmers' debts and pushed the value of the dollar upwards, making U.S. goods less attractive on foreign markets. In the middle of these actions, outgoing President Jimmy Carter placed an embargo on the Soviet Union for its invasion of Afghanistan. The severe decline in demand exacerbated by federal political moves created a deep agricultural depression that lasted for years. By 1986 the value of farm products were at the lowest levels since the Great Depression. Many American farmers were plunged into debt and were bought out by wealthier farmers, who had used wiser strategies or who had been lucky.[49]

The hastening of the consolidation of America's family farms into large-scale factory farms occurred alongside a groundswell of sentiment against agribusiness practices. Environmentalists and others began voicing concerns about agrichemical pollution, the depletion of natural resources, and the sustainability of some farming practices. The science that had created fertilizers and pesticides came under attack when it was revealed that many of the chemicals used by farmers were concentrating in the soil and water on and near farms because they did not break down naturally and because farmers were using large quantities of them. Nitrogen, from fertilizers, was found in high concentrations in some areas of the Midwest and was linked to cancer and birth defects. A pesticide agent, DDT, was proved to build up in plants and animals, interfering with reproduction and thereby exterminating wildlife.[50]

Environmentalists and political activists also began to worry about natural resources, such as petroleum, water, and soil, that were disappearing due to aggressive agricultural practices. The United States was dependent on foreign countries for its petroleum supply, an urgent fact during the petroleum shortage in the 1970s when OPEC (Organization of Petroleum Exporting Countries), a foreign oil cartel, manipulated the supply of petroleum so that it could raise prices dramatically. Twentieth-century American farmers relied increasingly on petroleum to run farm machinery, thereby increasing America's dependency on and depleting reserves of oil. At the same time in some areas of the United States, such as the central and southern Plains regions, irrigation depleted water reserves to dangerously low levels by the 1990s. Just as supplies of water and petroleum can be overspent, soil can be depleted through erosion and chemicals that kill off microorganisms necessary for breaking down organic matter. Environmentalists and activists worried about losing completely and irrevocably these nonrenewable resources.[51]

Interested individuals also began to question the wisdom of growing only the highly bred crops and livestock that most American farmers were raising. If a disease mutated so that the few strains of corn being raised were no longer immune to it, it could wipe out the entire crop. In the same vein, insects and weeds could become immune to the usual pesticides and wipe out crops or livestock. These arguments, coupled with environmental concerns, caused some Americans to begin questioning the sustainability of agribusiness practices.[52]

When federal farm policy came up for discussion during the 1980s, the traditional political farm network, which had at its core established farm organizations and the USDA, were joined by lobbyists with far different agendas. Environmentalists and activists concerned with pollution, nonrenewable resources, and the sustainability of agribusiness practices were flanked by unions that were angry about the treatment of farmworkers and their families, and academics who claimed that the government system of price and income supports had created a class of impoverished, landless, rural Americans. The population of the agricultural sector of the economy was so small by this time—2.5 percent of the total population in 1981—that the agricultural network was forced to consider the issues raised by the outside groups. Legislation began to reflect these issues, freezing or lowering quotas and/or payments for specific commodities, providing some environmental protections, and supporting research on low-technology sustainable agriculture.[53]

IHS BASS PHOTO CO. COLLECTION, #42574

Sprinkler watering system, 1915

Many farmers made alternative decisions after the boom of the 1970s, too. In order to retain their property, some farmers took full-time jobs off the farm and began farming only in their spare time. Thus, by 1991 nearly half of America's farmers had become "weekend farmers" with agricultural operations grossing less than $10,000 annually. Some farmers tried new commercial crops, such as the oil-seed crambe or Christmas trees, while others raised organic fruits and vegetables without chemicals for a growing market of "health-conscious Americans." Some farmers moved from grain crops to livestock, and others raised exotic animals like ostriches. At the same time, some farmers began to plan their farms somewhat less on profitability in terms of yield and more on sustainability in terms of nurturing the land and wildlife and the people whom they fed. In this way, many farmers became allies with the environmentalists and other nonfarm lobbyists.[54]

By 1990 only 1.8 percent of Americans still lived on farms and only

The woods were so dense that the trees were blazed for directions and paths. In the day light and after dark grandmother [Mary (Richards) Walker] would call to grandfather [Hugh Walker] to guide him to the house.

Alta Shellabarger
Delaware County

IHS BASS PHOTO CO. COLLECTION, #51326F

Sheep on Bray Farm, 1920

Opposite: Turkey farm, West Newton, 1934

half of the nation's farms were viable operations. Farmers had become a tiny minority whose subsidies had been trimmed by the federal government for a decade. Finally, Congress passed the Federal Agriculture Improvement and Reform Act of 1996, which established a plan whereby the government would end most subsidies for agriculture and return it to the marketplace within a few years. However, after passage of the bill, the federal government was forced to write emergency farm relief legislation in 1998 and 1999 as farmers' incomes predictably fell. Thereafter, George W. Bush's administration backed a new farm bill in 2002 that would reverse the 1996 bill.[55]

A new century in farming is bringing a new irony—small, independent farmers called "family farmers" are fighting against government agricultural programs their forebears in the farming industry helped to create. More than 125 years after farmers began to lobby the government, many family farmers perceive that the traditional twentieth-century federal farm programs are speeding their demise. Heated rhetoric surrounding the Farm Security and Rural Investment Act of 2002, which was passed by Congress in May 2002, highlights the schism that has developed between family farmers and agribusiness. By overturning the 1996 legislation, the 2002 bill returned to the philosophy of government intervention in the agricultural market in order to provide stable prices and higher incomes for farmers who produce specific crops and livestock. Heated press releases by numerous family farm coalition groups called for a rejection of the bill while simultaneously backing provisions in the bill aimed at conserving land and protecting wildlife, supplying food stamps for immigrant farm laborers, working to create renewable energy sources on farms, and mandating country of origin labels for food products. In fact it was the traditional portions of the bill that raised the farm coalitions' ire. Press releases by these groups accused the 2002 farm bill of supporting large factory farms because it subsidized the major grain crops and livestock, as well as allowing meat packers to own livestock. The coalitions asserted that the bill did little or nothing to support the diverse types of crops grown by smaller farmers, and they charged that the bill would stimulate the growth of new farm conglomerates at the expense of family farmers.[56]

Will the experience of being a small, independent farmer become extinct this century? Two hundred years of Indiana agricultural history indicate that it may. Except for the handful of Miami Indians in northern Indiana who managed to farm throughout the nineteenth and early twentieth centuries, Hoosiers have not actually lived a subsistence existence since the pioneer days. The pioneers' choices indicate that they did not move to the wilderness to live a meager existence; they came to

IHS BASS PHOTO CO. COLLECTION, #229836F

find a better way of life. Independent farming was supposed to make their lives richer in financial ways as well as in ways of personal freedom, space, and opportunity. Thus farmers readily brought with them the seeds of capitalism and industry that would destroy their initial subsistence way of life. Pioneers and those who followed built roadways and waterways to connect their farms to larger marketplaces, and they enthusiastically adopted new technologies in order to farm more productively. They bought and improved as much land as they could. Their sons and daughters and grandchildren learned about the agricultural experiments being conducted at Purdue University and elsewhere and embraced new ways of farming as they were able to do so. Their descendents continued the process of learning; improved their land, crops, animals, and processes; and built larger farming operations through boom times and bust. Those who survived as farmers enjoyed the parity granted by the federal government in the post–World War II era.

All the while, the very choices farmers were making helped to create, develop, and strengthen an agricultural industry in the country. Right away, the industry started gobbling up the poorest farmers, such as blacks and people of color; those who lost money through bad weather, bad investments, ill health, and other tragedies; children who were disinherited as elder siblings gained the family farms; and farmers who lost the business competition at crucial times such as the Civil War era, the Great Depression, and the end of the twentieth century. While the American agricultural industry as a whole made miraculous gains in production per unit of input, while the American farmer in the aggregate became capable of feeding the entire country—some suggest the entire world—small farmers became farm laborers, farm renters, part-time farmers, organic farmers, or exotic farmers. Many more farmers left the business altogether. Indeed, many Americans believe that the very success of American agribusiness—a success fueled by more than 125 years of lobbying by farmers with all sizes of operations—is steadily ringing in the end of the family farmer. According to this perspective, unless the American people and their government determine to make new choices in the face of global competition, environmental threats, and shrinking resources, dwindling prospects for family farmers seem clear.

I do not know how much [land] was entered at that time, [August 1811] but as he [John Ewbank] moved out to the wilderness in November of the same year and wintered in an abandoned squatters cabin, in the creek bottom between where the stone church stands and where the big hollow comes down through Albert Sumsers' land, while he cut and hauled logs and split out clapboard for the roof and built his cabin in the dooryard of this house before spring, I suppose this land was his original purchase.

Leobo J. Ewbank
Dearborn County

IHS INDIANA EXTENSION HOMEMAKERS COLLECTION

Blackford County neighbors come to help with the plowing

Notes:

1. Dave O. Thompson Sr. and William L. Madigan, *One Hundred and Fifty Years of Indiana Agriculture* (Indianapolis: Indiana Historical Bureau, 1969), 37.
2. James H. Madison, *The Indiana Way: A State History* (Bloomington and Indianapolis: Indiana University Press and Indiana Historical Society, 1986), 28.
3. Stewart Rafert, *The Miami Indians of Indiana: A Persistent People, 1654–1994* (Indianapolis: Indiana Historical Society, 1996), 64–65.
4. Ibid., 41.
5. Ibid., 79.
6. James M. Berquist, "Tracing the Origins of a Midwestern Culture: The Case of Central Indiana," *Indiana Magazine of History* 77 (March 1981): 9–26; Gregory S. Rose, "Hoosier Origins: The Nativity of Indiana's United States–Born Population in 1850," *Indiana Magazine of History* 81 (September 1985): 225–26.
7. Rafert, *Miami Indians of Indiana*, 95–118, 124, 141–49.
8. Henry Clyde Hubbart, *The Older Middle West, 1840–1880: Its Social, Economic and Political Life and Sectional Tendencies Before, During and After the Civil War* (New York: D. Appleton-Century Co., 1936), 91–92; Willard W. Cochrane, *The Development of American Agriculture: A Historical Analysis* (Minneapolis: University of Minnesota Press, 1979), 61.
9. Logan Esarey, *A History of Indiana* (Indianapolis: Hoosier Heritage Press, 1970), 606; Thompson and Madigan, *One Hundred and Fifty Years of Indiana Agriculture*, 17, 19; Madison, *Indiana Way*, 326.
10. Stephen A. Vincent, *Southern Seed, Northern Soil: African-American Farm Communities in the Midwest, 1765–1900* (Bloomington and Indianapolis: Indiana University Press, 1999), xi–xiii. For an example of an African-American pioneer community in Indiana, see Judith Q. McMullen, "African-American Pioneers and their Descendants in Harrison County," *The Hoosier Genealogist* 41 (September 2001): 179–84.
11. Vincent, *Southern Seed, Northern Soil*, xiii–xiv.
12. Cochrane, *Development of American Agriculture*, 52–55.
13. Randy K. Mills, "'Placing Family History in a Larger Historical Context' or 'How to Dramatize Your Family Stories'" (paper presented at the Midwestern Roots: Family History and Genealogy Conference, Indiana Historical Society, Indianapolis, 13–14 July 2001), 13–18; Thompson and Madigan, *One Hundred and Fifty Years of Indiana Agriculture*, 4.
14. Mills, "'Placing Family History in a Larger Historical Context,'" 13–18; Cochrane, *Development of American Agriculture*, 51–52, 71–72; Thompson and Madigan, *One Hundred and Fifty Years of Indiana Agriculture*, 7.
15. Cochrane, *Development of American Agriculture*, 67–68; Percy W. Blandford, *Old Farm Tools and Machinery: An Illustrated History* (Fort Lauderdale, Fla.: Gale Research, 1976), 43–60; Thompson and Madigan, *One Hundred and Fifty Years of Indiana Agriculture*, 4.
16. Cochrane, *Development of American Agriculture*, 67–69; Blandford, *Old Farm Tools and Machinery*: harrows, 102–107; cultivators, 107–10; cradles, 116; reapers, 117–22; threshers, 125–33.
17. Thompson and Madigan, *One Hundred and Fifty Years of Indiana Agriculture*, 17–18, 21, 27–28; Cochrane, *Development of American Agriculture*, 71–73.
18. Cochrane, *Development of American Agriculture*, 65–66; Thompson and Madigan, *One Hundred and Fifty Years of Indiana Agriculture*, 42. For the period 1800 to 1860, see also David B. Danbom, *Born in the Country: A History of Rural America* (Baltimore: Johns Hopkins University Press, 1995), 65–108.
19. Cochrane, *Development of American Agriculture*, 89–95; Susan Sessions Rugh, *Our Common Country: Family Farming, Culture, and Community in the Nineteenth-Century Midwest* (Bloomington and Indianapolis: Indiana University Press, 2001), 134–37.
20. Thompson and Madigan, *One Hundred and Fifty Years of Indiana Agriculture*, 19; Vincent, *Southern Seed, Northern Soil*, xii–xiv; Rafert, *Miami Indians of Indiana*, 151–68.
21. Thompson and Madigan, *One Hundred and Fifty Years of Indiana Agriculture*, 19–24; Cochrane, *Development of American Agriculture*, 95–96.
22. Cochrane, *Development of American Agriculture*, 96–106; Thompson and Madigan, *One Hundred and Fifty Years of Indiana Agriculture*, 21–22.
23. Thompson and Madigan, *One Hundred and Fifty Years of Indiana Agriculture*, 21, 29–30, 33–35. For the period 1870 to 1900, see also Danbom, *Born in the Country*, 132–60.
24. Thompson and Madigan, *One Hundred and Fifty Years of Indiana Agriculture*, 35; Cochrane, *Development of American Agriculture*, 243–44.

IHS BASS PHOTO CO. COLLECTION, #20107

Farmers market, 1910

Jeremiah Thompson purchased the land of James Benham for about $1.30 per Acre.

Jesse J. Bland
Sullivan County

A few years ago I made this barn smaller and the workmen used a double team of horses to pull down some of the supports, all complete logs of walnut or oak. Not a nail was found in the construction of the barn, still in use.

Elizabeth Asenath Hall
Hamilton County

25. Cochrane, *Development of American Agriculture*, 107–11.

26. George Brown Tindall and David E. Shi, *America: A Narrative History*, 2 vols., 4th ed. (New York: W. W. Norton and Co., 1984), 2:884–87.

27. Cochrane, *Development of American Agriculture*, 109–11.

28. Thompson and Madigan, *One Hundred and Fifty Years of Indiana Agriculture*, 48.

29. Cochrane, *Development of American Agriculture*, 100–101; Thompson and Madigan, *One Hundred and Fifty Years of Indiana Agriculture*, 49–50.

30. Cochrane, *Development of American Agriculture*, 100–101; Thompson and Madigan, *One Hundred and Fifty Years of Indiana Agriculture*, 48–50; Tindall and Shi, *America*, 2:877, 886; Vincent, *Southern Seed, Northern Soil*, xii; Rafert, *Miami Indians of Indiana*, 196–204. For the period 1900 to 1920, see also Danbom, *Born in the Country*, 161–84.

31. Thompson and Madigan, *One Hundred and Fifty Years of Indiana Agriculture*, 49–51.

32. Ibid., 25; Cochrane, *Development of American Agriculture*, 116–21.

33. Thompson and Madigan, *One Hundred and Fifty Years of Indiana Agriculture*, 25–26.

34. Cochrane, *Development of American Agriculture*, 112–13.

35. Thompson and Madigan, *One Hundred and Fifty Years of Indiana Agriculture*, 51; Madison, *Indiana Way*, 297–98. For the period 1920 to 1933, see also Danbom, *Born in the Country*, 185–205.

36. Cochrane, *Development of American Agriculture*, 140–41; Congressional Research Service, "Farm Commodity Legislation: Chronology, 1933–98," edited by Geoffrey S. Becker (Washington, D.C.: Library of Congress, 1999; published on the Internet by National Council for Science and the Environment, National Library for the Environment, www.cnie.org), 1–2. Brackets are inserted by the author and are not part of the original text.

37. Cochrane, *Development of American Agriculture*, 141–43; Congressional Research Service, "Farm Commodity Legislation," 2.

38. Thompson and Madigan, *One Hundred and Fifty Years of Indiana Agriculture*, 52; Cochrane, *Development of American Agriculture*, 227–28; National Archives, "Records of the Rural Electrification Administration [REA] (RG 221), 1934–73," *Guide to Federal Records in the National Archives of the United States* (published on the Internet by the National Archives, www.nara.gov), 221.1; Ralph Gann and Steve Wilson, "Number of Farms, 1900–1997" (n.p.: Indiana Agricultural Statistics Service, National Agricultural Statistics Service (NASS), U.S. Department of Agriculture, n.d., published on the Internet by NASS, www.nass.usda.gov), 2.

39. Cochrane, *Development of American Agriculture*, 143–44; Congressional Research Service, "Farm Commodity Legislation," 2–3.

40. Cochrane, *Development of American Agriculture*, 144, 378–95, and "About the Author" at back of book; Congressional Research Service, "Farm Commodity Legislation," 1–3. For the period 1933 to 1945, see also Danbom, *Born in the Country*, 206–32.

41. Thompson and Madigan, *One Hundred and Fifty Years of Indiana Agriculture*, 37–38, 53, 55; Tindall and Shi, *America*, 2:A36. Brackets are inserted by the author and are not part of the original text.

42. Thompson and Madigan, *One Hundred and Fifty Years of Indiana Agriculture*, 53–54; Cochrane, *Development of American Agriculture*, 228.

43. Thompson and Madigan, *One Hundred and Fifty Years of Indiana Agriculture*, 39–42, 45–46, 54; Cochrane, *Development of American Agriculture*, 201–2.

44. Thompson and Madigan, *One Hundred and Fifty Years of Indiana Agriculture*, 39–44, 54–55. For the period 1945 to 1970, see also Danbom, *Born in the Country*, 233–52.

45. Thompson and Madigan, *One Hundred and Fifty Years of Indiana Agriculture*, 55–57; Cochrane, *Development of American Agriculture*, 378–95.

46. Thompson and Madigan, *One Hundred and Fifty Years of Indiana Agriculture*, 55.

47. Danbom, *Born in the Country*, 253–55.

48. Ibid., 255–56.

49. Ibid., 262–63.

50. Ibid., 258–59.

51. Ibid., 259–61.

52. Ibid., 259–60.

53. Ibid., 256–62, 266; Congressional Research Service, "Farm Commodity Legislation," 4–5.

54. Danbom, *Born in the Country*, 266–68.

55. Ibid., 266; Congressional Research Service, "Farm Commodity Legislation," 5–6; Economic Research Service (ERS), "The 2002 Farm Bill: Provisions and Economic Implications" (Washington, D.C.: U.S. Department of Agriculture, 2002, published on the Internet by ERS, www.ers.usda.gov), 1.

56. For background to the 2002 farm bill, see *Food and Agricultural Policy: Taking Stock for the New Century* (Washington, D.C.: U.S. Department of Agriculture, 2001). For information about the 2002 farm bill, see ERS, "The 2002 Farm Bill." For examples of articles written by farm groups opposed to the farm bill, see the following press releases published in *Ag Observatory*, "What's Wrong with the Farm Bill" (Minneapolis: Institute for Agriculture and Trade Policy [IATP], 2002, published on the Internet by IATP, www.agobservatory.org): Bill Christison, "Family Farmers Express Strong Opposition to 2002 Farm Bill" (press release of National Family Farm Coalition, May 2002); "Campaign for Family Farms Calls for Rejection of Reported Farm Bill" (press release of Campaign for Family Farms [a coalition that includes Missouri Rural Crisis Center, Iowa Citizens for Community Improvement, Illinois Stewardship Alliance, and Land Stewardship Project], April 2002); "Corn Growers Give Farm Bill a Failing Grade" (press release for American Corn Growers Association, May 2002); "Midwestern Voters Support Conservation Funding for Agriculture" (press release of American Farmland Trust, October 2001); "WORC Applauds Food Labeling, but Raps the Rest of Farm Bill" (press release of Western Organization of Resource Councils, April 2002). Copies of the above press releases are in the possession of the author. See also John Boehner and Cal Dooley, "This Terrible Farm Bill," *Washington Post*, 2 May 2002.

Index to the Original Owners of the Centennial Farms

ANCESTOR	COUNTY	DATE	DESCENDANT	RELATIONSHIP
ABBOTT, William N.	Dearborn	1821	James Scott Abbott	grandson
ABEL, Homer	DeKalb	1836	Clyde E. Hart	grandson
ABERNATHY, Hugh	Union	1832	Mrs. Clint Bryson	great-granddaughter
ABSHEAR, Abraham	Wabash	1836	Nellie (Abshire) Flohr	great-granddaughter
ADAMS, John	Parke	1830	Mrs. Arthur O. Ramsey	great-great-granddaughter
ADAMSON, William	Lawrence	1837	Lisker L. Adamson	grandson
ADSIT, William	Delaware	1838	Karl D. Nottingham	grandson
AITKENS, Joseph	Clinton	1847	Nola E. (Aitkens) Denison	granddaughter
ALBIN, Morland	Elkhart	1837	David C. Albin	great-grandson
ALEXANDER, Samuel R.	Knox	1839	Charles R. Alexander	great-grandson
ALLEE, Pleasant	Putnam	1838	Enos E. Allee	grandson
ALLEN, Hugh	Wayne	1832	Hanford Cleveland	great-grandson
ALLEY, Jonathan	Bartholomew	1831	Herschel Alley Blades	great-grandson
ALLGEIER, Lorenz	Allen	ca. 1840	Georgian Allgeier	great-granddaughter
ALTER, John	Jasper	1849	Christmas Eads Alter	grandson
AMON, Adam	Rush	1833	Emma Hood*	great-grandniece
AMOS, Joseph J.	Rush	1839	Willard H. Amos	grandson
ANDERSON, James	Putnam	1831	Jesse Fay Anderson	great-grandson
ANDERSON, John	Madison	1823	Mrs. Ralph Williamson*	not given
ANDERSON, John	La Porte	1836	Mrs. Lyle H. Pitner	great-granddaughter
ANDERSON, William, Sr.	Dubois		Laura (Anderson) Harrison	great-granddaughter
APPLE, Jessee	Orange	1841	Orpheus Apple	grandson
APPLEMAN, Leonard	LaGrange	183?	Cecil Appleman	great-great-grandson
ARBUCKLE, John	Jefferson	1847	Dallas Hardy	great-grandson
ARCHER, Robert	Gibson	1836	Charles A. Miller	grandson
ARMITAGE, Seth	Jay	1841	Earl F. Miller	grandson
ARMSTRONG, John	Washington		J. Beech Armstrong	grandson
ARNOLD, John	Rush	1820	Harvey Arnold	great-grandson

Compiled by

Ruth Dorrel

This home was assembled in [Covington,] Indiana, and floated down the Wabash River where it was erected on a spot [William White] chose on this land.

Chas. Morey White
Vermillion County

Capitol dairy wagon, 1942

ANCESTOR	COUNTY	DATE	DESCENDANT	RELATIONSHIP
ART, Thomas	Vigo	1835	Ella Smith	great-granddaughter
			Tom Smith	great-grandson
			Welton M. Smith	great-grandson
ASHBY, William	Pike	ca. 1854	Laura Jackson	great-granddaughter
ATHERTON, Stout	Fayette	1833	Mrs. James Riggs	granddaughter
ATKINS, Joseph	Tippecanoe	1839	Jesse Charles Andrew	grandson
ATKINS, Thomas	Rush	1837	Herman Atkins	great-grandson
ATKINSON, James	Henry	1832	Elbert Riggs	great-great-grandson
AYLESWORTH, Giles	Porter	184?	James P. Aylesworth	great-grandson
AYRES, Azariah	Gibson	1823	Clarence LaGrange	great-great-grandson
BACON, Socrates	Allen	1838	Harvey E. Bacon	great-grandson
BAILEY, Martin	Marshall	1841	Mrs. Harmon Leffert	great-granddaughter
BAITY, George	Hancock	1834	Orville E. Baity*	grandnephew
BAKER, Abraham	Kosciusko	1847	Beryl Jefferies	granddaughter
BAKES, John	Switzerland	1845	Bernice (Bakes) Poston	granddaughter
BALLARD, Taylor, Sr.	Johnson	1833	Lem B. Tilson	grandson
BANGS, Heman	DeKalb	1839	Owen R. Bangs	grandson
BANTA, Abraham	Cass	1836	Caleb Banta	grandson
BANTA, Beauford	Cass	1834	Caleb Banta	son
BANTA, John Peter	Johnson	1826	Frank C. Banta	great-grandson

ANCESTOR	COUNTY	DATE	DESCENDANT	RELATIONSHIP
BARGER, Michael	Marshall	1838	Arthur C. Berger	great-grandson
			Valera G. (Berger) Tedrow	great-granddaughter
BARKER, Thomas L.	Hamilton	1849	Sibil (Barker) Harvey	granddaughter
BARLOW, Jacob	Johnson	1832	Byron Runkle	great-grandson
			Lewis Runkle	great-grandson
			Monta Runkle	great-granddaughter
BARNARD, William	La Porte	1835	Charles N. Barnard	great-grandson
BARNARD, William	Porter	1835/37	Charles N. Barnard	great-grandson
			Harold Barnard	not given
BARNES, C. B.	Decatur	1834	Cortez Barnes	great-grandson
BARNETT, Valentine	Miami	1846	Claude A. Dice	great-grandson
BARRETT, George	Posey	1817	Hortense Barrett	great-granddaughter
			Olive Barrett	great-granddaughter
BARRETT, John	Hancock		John R. Trees	great-grandson
BARROW, Richard	Allen	1835/36	Helen Harper	granddaughter
			John Harper	grandson
			Maud Harper	granddaughter
			Melissa Harper	granddaughter
BARTER, William	Posey	1836	Job Barter	grandson
BARTHOLOMEW, Joseph	Porter	1835	Gerald Bartholomew	grandson
BARTHOLOMEW, Noah	Marshall	1842	George B. See	great-grandson
			Lloyd See	great-grandson

IHS BASS PHOTO CO. COLLECTION, #1083263

This land, which David Stalker bought, was previously a part of a school land section which had been donated by an act of Congress in May 1822 for the use of schools in Clarks Grant. Therefore, Mr. Stalker was the first private owner of this piece of land.

Ellis Barker
Hamilton County

In 1821 September 19, [William Tyner] entered eighty acres of land in Blue river township, being the third entry made in the county. [He] was one of the staunch pioneers, coming into the county within two years from the first settlement made by the paleface.

Nellie Shadley
Hancock County

ANCESTOR	COUNTY	DATE	DESCENDANT	RELATIONSHIP
BATSON, Esther	Henry	1847	Lamont O'Harra	great-great-grandson
BAUMAN, Levi	Fulton	1845	Vallie A. Maudlin	granddaughter
BEALS, Thomas	Hamilton	1848	Waldo Beals	grandson
BEAN, Pleasant D.	Harrison	1833	Walter D. Bean	grandson
BECK, Elizabeth	Montgomery	1840	Jesse Caster	grandson
BECK, John	Tipton	1837	Charles H. Beck	grandson
BECKMAN, Henry	Franklin	1835	Ralph Beckman	grandson
BECKNER, Jeremiah	Rush	1826	Mrs. John A. Nelson	granddaughter
BEESON, Benjamin	Wayne	1814	Alice (Beeson) Kniese	great-granddaughter
BEESON, David	Wayne	1830	Olive R. Beeson*	granddaughter-in-law
BELL, John	Union	1825	J. Arthur Bell	great-grandson
			Mary K. Witter	great-granddaughter
BELL, John	Henry	1830	Ida Kendall	great-great-granddaughter
BELL, John	Rush	1831	Fred B. Bell	grandson
BELL, William	Henry	1823	Charles Bell	grandson
			Floyd Bell	great-grandson
BENDER, John	Spencer	1839	Noah N. Bender	great-great-grandson
BENHAM, James	Ripley	1815	Dora (McCoy) Maxwell	great-great-granddaughter
BENICIT, Abel	Shelby	1829	Mrs. J. J. (Bennett) Kemper	great-granddaughter
BENNETT, Jesse	Rush	1822/45	Lillian O. Bennett	great-granddaughter
			Ruth E. Bennett	great-granddaughter
BENSON, David	Posey	1813	Zack F. Smith	grandson
BENSON, William	Gibson	1815	Lillie (Benson) Smith	granddaughter
BENTLEY, Elisha	Vigo	1830	Frank Bentley	great-grandson
BERGEN, Peter	Johnson	1828	Paul V. Covert	great-grandson
BERKSHIRE, John	Harrison	1839	Caddie (Berkshire) Benz	great-granddaughter
BERRY, Jonathan	Grant	1846	Abbie S. Weaver	granddaughter
BERRY, Michael	Clark	1819	Roy J. Ross	great-great-grandson
BETTERTON, William	Madison	1837	Estella J. Hull	great-granddaughter
			Loren E. McCarty	great-grandson
BETTS, Cyrus	Switzerland	1841	Mrs. Dawson W. Bakes	great-granddaughter
BEUOY, David	Delaware		Charles T. Beuoy	grandson
BEVER, Michael	Rush	1821	Jessie (Bever) Cameron	great-granddaughter
BICKWERMENT, Herman	Dubois	1848	Charles F. Pund	great-grandson
BIDDLE, Caleb	Madison	1830	Glen Elsbury	great-grandson
			Vergie Tweedy	great-granddaughter
BIDDLE, William	Fulton	1837	Elza A. Decker	grandson

IHS INDIANA EXTENSION HOMEMAKERS COLLECTION

Threshing with mules.
Grant County

This farm has now seen the birth of the Fourth generation of descendants of George W. Ferguson, who came from Erie, Penn. to this section looking for a desirable place for him & his family to locate . . . what is now & has been ever since, "the Ferguson Farm"—the deed transfer was made July 16- 1836. . . . In 1855 G. W. Ferguson built a big 15-room house on it but the original old 1836 house still stands there.

Mary Jane Ferguson
LaGrange County

ANCESTOR	COUNTY	DATE	DESCENDANT	RELATIONSHIP
BIGGS, George Crist	Sullivan	1845	Cora Jane Austin	granddaughter
			Nancy Margaret Lloyd	granddaughter
BILLINGSLY, Samuel	Johnson	1834	Bynum L. Billingsly	grandson
BILSLAND, John	Fountain	1824	Mrs. Samuel Van Dorn	great-granddaughter
BISHOP, Cyrus W.	Vigo	1838	Sue E. (Bishop) Krog	granddaughter
BISHOP, James Harvey	Kosciusko	1832	James E. Bishop	grandson
BISHOP, Joel	Jefferson	1838	Agnes L. Bishop	granddaughter
BITNER, Stephen	Rush	1832	Arthur Bitner	grandson
BLACK, Amos	Noble	1845	Bess (Black) Benton	granddaughter
			Martha (Black) Homsher	granddaughter
BLACK, William	Delaware	1834	William H. Black	son
BLACKLIDGE, John	Rush		Allen H. Blacklidge	great-grandson
BLACKMAN, Almond	LaGrange	1839	Royal Blackman	grandson
BLAND, James	Jefferson	1814	Marion C. Bland	great-grandson
BLUE, Uriah	Carroll	1848	Guthrie Blue	great-grandson
			Susan Blue	great-granddaughter
BLUNK, Goldsbury	Morgan	1828	Raymond Dale Blunk	great-great-grandson
			Ruth Sims	great-granddaughter
			Mrs. Florence Blunk*	granddaughter-in-law
BLY, Jeremiah	Randolph	1844	Mary E. Bly*	daughter-in-law
BOHNER, William	Ripley	1846	Lillie Bohner	granddaughter
BOLINGER, Elijah	Madison	1835	Rolland E. Bollinger	grandson

My grandfather [Peter Miller] was 19 years old when this farm was entered from the government, was married, hence making himself sure of the land he wanted, but not being of age, had to enter the farm in the name of an older person, his brother [Henry Miller].

Elizabeth Quebbeman
Harrison County

ANCESTOR	COUNTY	DATE	DESCENDANT	RELATIONSHIP
BOLYARD, Amos	Allen	1844	Andrew Bolyard	not given
BOND, Jesse	Wayne	1814	Edith (Bond) Morgan	great-granddaughter
BOND, William	Miami	1844	Edward R. Coppock	great-grandson
BONE, Adam	Carroll	1842	Robert Wingard	great-great-grandson
BONEWITZ, ——	Huntington	1837	Dale Bonewitz	son
BOTKIN, Hugh	Randolph	1817	John W. Botkin	grandson
BOUSE, Aaron	Noble	1846	Mabel (Bouse) Hontz	granddaughter
BOWELL, Absalem Carr	La Porte	1847	Lena M. Benjamin	granddaughter
			E. Carr Bowell	grandson
			Zayda G. Bowell	granddaughter
BOWEN, Ephraim	Randolph	1838	Florence Ruth (Bowen) Taylor	great-granddaughter
BOWERS, John	Marion	1846	Edgar H. Bowers	great-grandson
BOWMAN, Daniel	Henry	1832	William R. Bowman	great-grandson
BOWMAN, Henry	Boone	1834	Carl C. Bowman	great-grandson
BOWMAN, John	Miami	1846	Edna M. Bowman	great-granddaughter
BOYER, Jacob	Clark	1831/35	Omer C. Boyer	grandson
BOZWORTH, William	Clinton	1847	Edith M. Curts	granddaughter
BRADEN, Burr	Clinton	1830	William R. Braden	great-grandson
BRADLEY, Henry	Johnson	1845	William B. Oliver	great-grandson
BRADY, Absalom	Spencer		Howard Parsley	great-grandson
BRANDENBURG, John	Harrison	1835	Roy Brandenburg	grandson

IHS BASS PHOTO CO. COLLECTION, #229423F

Fruit and vegetable farm, 1934

IHS BASS PHOTO CO. COLLECTION, #103029F41

ANCESTOR	COUNTY	DATE	DESCENDANT	RELATIONSHIP
BRANDT, John H.	Ripley	1846	Louis Brandt	grandson
BRATTON, William	Montgomery	1823	Maud J. (Bratton) Chesterson	great-granddaughter
BRAY, Edward	Hamilton	1835	Alma Bray	great-granddaughter
			Herbert E. Bray	great-grandson
			Hazel (Bray) Richards	great-granddaughter
BREECE, James	Posey	1837	Pearl (Breece) Allyn*	grandniece
BREEDEN, Sterling	Randolph	1834	Hattie Breeden*	granddaughter-in-law
BREES, David	Posey	1835	Olive (Breece) Allyn*	niece
			Pearl (Breece) Allyn*	niece
BRENTLINGER, George	Knox	1841	Maude (Brentlinger) Hohn	granddaughter
BRETZ, Jacob, Sr.	Dubois	1837	Benjamin B. Bretz	great-grandson
BREWER, David D.	Johnson	1834	Guy D. Brewer	grandson
			Nellie M. Brewer	granddaughter
			T. Smith Brewer	grandson
BREWER, John D.	Johnson	1832	Mino Dickson	grandson
BREWER, Samuel	Jackson	1838	S. J. Brewer	grandson
BRIER, Isaac	Warren	1830	William E. Brier	great-grandson
BRIGHT, Levi	Decatur	1840	Roscoe Bright	grandson
BRISCOE, William	Warrick	1818	Herman S. Collins	great-great-grandson
BRITTS, Samuel	Montgomery	1831	H. Clay Britts	great-grandson
BROCKSMITH, John	Knox		Gardner Brocksmith	great-great-grandson
BROOKS, Humphrey	Jefferson	1816	Arthur Brooks	grandson

John Miller fought in the War of 1812. Came from Virginia and settled or purchased this farm. He went back to Virginia mostly by ox-cart and brought my grand mother to Indiana. Farm has never been out of the Miller Name since 1825.

Burke H. Miller
Hendricks County

Joseph Koons built 1st Log House without the use of any iron. He was a big, strong man. Would walk to Richmond and carry back 75 lbs of material. Would try a sled of [wood] himself before Hitching his Horse to it. The Koons still have a gourd, which Joseph received from Indians in exchange for corn.

Benjamin Koons
Henry County

William F. Johnson Lumber Company, chicken yard, 1926

ANCESTOR	COUNTY	DATE	DESCENDANT	RELATIONSHIP
BROOKS, James	Jefferson	1811	Viola (McKay) Coleman	granddaughter
			Fred H. McKay	grandson
BROOKS, William	Jefferson	1830	Harry H. Brooks	great-great-grandson
BROWN, Daniel	Carroll	1840	Lottie (Brown) Cooke	granddaughter
BROWN, Francis Marion	Owen	1844	Ralph E. Brown	grandson
BROWN, Francis O.	Monroe	1830	Donna Brown	granddaughter
			Ida L. Buzzaird	granddaughter
			Robert L. Buzzaird	great-grandson
			Blanche (Brown) Dodds	granddaughter
BROWN, John	Montgomery	1829	Maud M. Brown*	granddaughter-in-law
BROWN, John D.	Howard	1847	James M. Brown	grandson
BROWN, Johnston	Jefferson	1833	Martin Brown	not given
BROWN, Joseph	Wayne	1818	Floyd E. Brown	grandson
BROWN, Michael	Fayette	1814	Lloyd Nickels	great-great-grandson
BROWN, Robert	Marion	1822	Russ R. Reading	grandson
BROWN, Russel	LaGrange	1836	Laura (Brown) Talmage	granddaughter
BROWN, Samuel	Parke	1825	Martha Brown	granddaughter
BROWN, Samuel	Putnam	1835	Livia M. Ashby	great-granddaughter
BRUCE, Amour	Dearborn	1797	Mrs. Adam Bruce*	granddaughter-in-law
			Laura Dennerline	great-granddaughter
			Mrs. J. Parks	granddaughter
			Mabel Warner	granddaughter

M. Rumely Company, Advance Rumely Separator

IHS DAVID PEAT COLLECTION

I am glad our old ancestor [David Vogler] had enough sense to leave Germany and come to America, and raise a large brood of Americans for America . . . A man said he did not care what his grandfather was, he was more interested in what his grandson was going to be.

Lucy Vogler
Bartholomew County

ANCESTOR	COUNTY	DATE	DESCENDANT	RELATIONSHIP
BRUCE, James	Harrison	1807	Edna Bruce	not given
			Georgia Bruce	not given
BRUCE, William	Knox	1805	Cora (McClure) Goodman	great-granddaughter
BRUMFIELD, Thomas	Wayne	1826	Glen Brumfield	great-great-grandson
BRUNCK, Peter F.	Fulton	1839	Gladys (Brunk) Richards	granddaughter
BRUNEMER, Conrad	Johnson	1831	Grace C. Hilderbrand	great-granddaughter
BRUSS, William	Huntington	1844	Guy William Bruss	grandson
BRYANT, Andrew	Parke	1847	Otto Bryant	grandson
BUCHANAN, George	Ripley	1813	George T. Buchanan	great-grandson
BUCHANAN, James	Ripley	1816	John Maxwell	great-grandson
BUCHANAN, James	Parke	1820	Warren Buchanan	great-great-grandson
			Mildred Hall	great-great-granddaughter
BUCHANAN, James	Cass	1839	Mrs. Clyde Davidson	granddaughter
BUCHANAN, John	Fayette	1821	Anna (Smith) Grey°	grandniece
BUCHANAN, William	Montgomery	1842	Goldie (Buchanan) Williamson	granddaughter
BUCKLEY, Dennis	Lake		Charles A. Buckley Jr.	great-grandson
			Gerald Albert Buckley	grandson
BUCKNER, Avery M.	Johnson	1835	Harley Buckner	great-grandson
			Walter Buckner	great-grandson
			Charles M. Green	great-grandson
BUENTE, George Ludwig	Vanderburgh	1843	Wilbert C. Buente	great-great-grandson
			Wilburn H. Buente	great-great-grandson

I remember of hearing my mother tell how Grandpa [John Stelzer] walked from here to Ft. Wayne a distance of 50 miles to make the deed and pay for the land. He walked fast all the way there for he was afraid someone might rob him of the money he had.

Mrs. Fedalis Wagner
Jay County

ANCESTOR	COUNTY	DATE	DESCENDANT	RELATIONSHIP
BURCH, Charles	Monroe	1835	Mrs. G. Manson Mood	granddaughter
BURGESS, James	Delaware	1838	James W. Burgess	great-grandson
			Mary Burgess	great-granddaughter
			Rennie Burgess	great-granddaughter
			Carrie (Burgess) Harvey	great-granddaughter
BURGESS, John	Delaware	1840	Samuel A. Burgess	great-grandson
BURGET, Aaron	Johnson	1827	Donald B. Land Sr.	great-grandson
			Lewis B. Richardson	great-grandson
BURRIS, Matthew	Delaware	1838	Harry A. McColm	great-grandson
			Joseph H. McColm	great-grandson
			Ralph W. McColm	great-grandson
BURTON, John	Monroe	1831	Mrs. Dewey D. Douglas	great-great-granddaughter
BURTON, Robert	Gibson	1846	Ronald C. Burton	grandson
BUSSELL, John	Henry	before 1847	Oliver E. Bussell	grandson
BUTLER, Justus	Steuben	1836	William Justus Butler	grandson
BUTLER, Lemuel	Montgomery	1845	Gladys (Butler) Abbott	great-granddaughter
BUTT, Thomas	Randolph	1838	Anna L. Butts	granddaughter
BUTT, William	Allen	1832	J. F. Butt	great-grandson
BUTTERFIELD, Luke G.	Marion	1865	Amos A. Butterfield	son
BUTTLER, Hiram	Switzerland		William Bradford	grandson
BYALL, James	Allen	1835	Gladys A. Fritz	great-granddaughter

M. Rumely traction engine, Advance Rumely Separator

IHS DAVID PEAT COLLECTION

ANCESTOR	COUNTY	DATE	DESCENDANT	RELATIONSHIP
BYRD, Abraham	Montgomery	1827	Alton Byrd	great-great-grandson
			Hattie Byrd*	great-granddaughter-in-law
			J. Thomas Byrd	great-great-grandson
BYRKET, Noah	Boone	1831	Bessie M. Shelburne	granddaughter
CAIN, John	Grant	1835	Walter Cain	grandson
CALDWELL, Alexander	Boone	1839	Ivy Caldwell	granddaughter
CALDWELL, Joseph I.	Fayette		Eva Caldwell	great-great-granddaughter
CALLAHAN, Ami	LaGrange	1846	Lulu Whitmore	granddaughter
CAMPBELL, Adam S.	Porter	1833	Mary Edna (Campbell) Cain	great-granddaughter
CAMPBELL, Samuel	Fountain	1827	Jennie C. Glascock	great-granddaughter
CAMPBELL, Samuel G.	Delaware	1832	Anna (Campbell) Powers	granddaughter
CAMPBELL, Williamson	Boone	1844	Elza B. Campbell	great-grandson
CANINE, Ralph	Montgomery	1827/36	Edwin N. Canine	great-grandson
			Foster Oldshue	great-grandson
CAPPER, Martin	Decatur	1835	Earl W. Capper Sr.	great-grandson
CAREY, Zenas	Hamilton	1835	Lowell Carey	great-grandson
			Mary M. Carey	granddaughter
CARLTON, William	Madison		Nora Swart*	granddaughter-in-law
CARMONY, John	Shelby	1836	Arthur Carmony	great-grandson
CARR, Samuel A.	Scott	1843	Irene J. Gibson	great-granddaughter
CARROLL, J. H.	Decatur	1845	W. C. Carroll	grandson
CARSON, Joseph	Rush	1825	Vista Newkirk	great-granddaughter
CARSON, William	Hamilton	1839	Charles Carson	grandson
			Mrs. E. J. Nance	great-granddaughter
CARTER, Nathan	Bartholomew	1820	Mrs. Silvan E. Carter*	great-granddaughter-in-law
CARTMEL, Joel H.	Decatur	1845	Mrs. Edgar E. Hite	granddaughter
CARVER, Elijah	Fayette	1826	Lucille B. French	great-great-great-granddaughter
CASE, Caleb	DeKalb	1843	Sylvia Ginther	granddaughter
CASTER, John	Montgomery	1846	Jesse Caster	grandson
CASTERLINE, Ira	Blackford	1836	Ida (Casterline) Rice	granddaughter
CATT, Phillip, Jr.	Pike	1835	Charles Catt	great-grandson
CAYLER, Samuel	Randolph	1837	Samuel H. Cayler	grandson
CHAMBERS, Elijah	Owen	1818	Ivan Chambers	great-great-grandson
CHARLES, William	Orange	1836	William H. Charles	great-grandson
			Grace S. Charles*	granddaughter-in-law
CHARLEY, Sarah B. (Hayden)	Harrison	1833	Otto Hottel	grandson
			Sally (Lockhart) Miles	granddaughter

This 120 [acres] was bought from the government so they would have the timber to make rails for fences by Sol. R. Eastman

Harold Eastman
Jennings County

In 1835 erected first mill for grinding corn. The first religious meeting in the township was held on this farm in a cabin erected for that purpose in 1839. First election held at house of John R. Emley [first owner] first Monday April 1838, with 8 qualified voters. The first corn and potatoes were planted by John R. Emley in 18[?].

Faith Emley Rahn
Huntington County

ANCESTOR	COUNTY	DATE	DESCENDANT	RELATIONSHIP
CHITWOOD, Daniel L.	Jefferson	1843	Nora H. Chitwood°	granddaughter-in-law
CHRISTY, James	Putnam	1826	H. C. Christy	grandson
CISLING, Henry S.	Warren	1833	J. K. Chambers	great-grandson
CLARK, Hamilton	Sullivan	1840	Laura (Jeffries) Johnson	granddaughter
CLARK, John R.	Johnson	1825	Ogle D. Clark°	nephew
CLARK, John Y.	LaGrange	1836	Grace (Clark) Babcock	granddaughter
			Hester (Clark) Malone	granddaughter
CLARK, Priscilla	Rush	1847	Clark Offutt	grandson
CLAYTON, Stephen	Randolph	1823	Stephen Elsworth Clayton	grandson
CLYBURN, Henly	La Porte	1829	Rollo Gardner°	stepson
COBLE, George	Marion	1829	Freda (Coble) Brammer	great-granddaughter
			Alice (Aston) Harvey	great-great-granddaughter
COE, William	Union	1804	Mrs. James Porter	great-granddaughter
COFFIN, Joseph J.	Hendricks	1835	Jonathan Lowe	grandson
COLE, Mathew, Jr.	Putnam	1837	Agnes M. King°	great-granddaughter-in-law
COLEMAN, John B.	Daviess	1836	Emery Lett	grandson
COLEMAN, Thomas	Grant	1829	Alva (Coleman) Mason	great-grandson
COLEMAN, Virgil	Ripley	1844	Charles S. Coleman	grandson
COLSON, Joel	Decatur	1837	James F. Colson	grandson
COLTRIN, William	Vigo	1819	Emma Coltrin	great-granddaughter
			Mary Coltrin	great-granddaughter
			Nina Keep	great-granddaughter
COLVIN, Andrew H.	Harrison	1832	Mildred (Colvin) Board	granddaughter
			John Abel Colvin	grandson
COMSTOCK, John	Wabash	1836	Charles Comstock	grandson
CONES, Washington	Hancock	1840	Ben Cones	grandson
CONOWAY, Robert	Dearborn	1811	Fleetwood Conoway	great-grandson
COPE, Jonathan, Sr.	Jefferson	1816	Taylor Cope	great-grandson
COPELAND, James	Jackson	1831	Gabie G. Turmail	great-granddaughter
COPELAND, John	Henry		Glen Copeland°	cousin
CORN, William	Montgomery	1833	George F. Frantz	great-grandson
CORNELIUS, George	Shelby	1834	Edward C. Gaines	great-grandson
CORNELL, John G.	Carroll	1836	Eva (Cornell) Brown	great-granddaughter
CORNTHWAITE, Guy	Parke	1835	Homer G. Cornthwaite	grandson
COSSEY, Peter	Vermillion	1844	Mabel (Cossey) LaTourette	granddaughter
COTTON, James	Fayette	1843	Wm. H. Coltrane	great-grandson
COULTER, John	Whitley	1836	Miriam (Coulter) McConnell	great-granddaughter
COX, Abner	Morgan	1822	Fisk Landers	great-great-grandson

IHS BASS PHOTO CO. COLLECTION, #213993F

Neimeyer Farm, 1929

Major Townsend's mother, "Plebe Collins" is the girl that hid from the Indians at the time of the "Pigeon Roost Massacre."

Roscoe E. Waltz
Johnson County

ANCESTOR	COUNTY	DATE	DESCENDANT	RELATIONSHIP
COX, Benjamin, Jr.	Randolph	1838	Edgar G. Cox	great-grandson
COX, Jeremiah I.	Wayne	1812	Mrs. Leslie R. Cook*	great-great-granddaughter-in-law
COX, Jonathan Piety	Knox	1836	Dobertine (Cox) Dees	great-granddaughter
CRAIG, Robert	Boone	1838	Lottie B. Johnson	granddaughter
CRANE, Asa	Jackson	1825	Mrs. Warren Skinner	granddaughter
CRAWFORD, James	Warren	1829	Verlie I. Blue	great-great-granddaughter
CRIPE, Daniel	Wabash	1836	David S. Cripe	great-grandson
CRISLER, Lewis	Shelby	1834	Ethel Crisler*	granddaughter-in-law
CRIST, ——	Jennings		Maud Crist	great-granddaughter
CROUSE, John W.	Tippecanoe	1831	P. L. Crouse	grandson
CROWL, Mickel	Kosciusko	1838	Alva Crowl	grandson
CROWNOVER, Benjamin	Wayne	1838	Frances Crownover	great-granddaughter
			James Ottus Crownover	great-grandson
			Mrs. Samuel Ottus Crownover*	granddaughter-in-law
CULLER, Jacob	Carroll	1845	Grace (Culler) Shaffer	granddaughter
CULLMAN, Peter	Perry	1847	Anna (Cullman) Zuspann	great-granddaughter
CUMMINGS, David	Lawrence	1847	Laura C. Brewer	granddaughter
			Jesse Cummings	grandson
CUMMINS, David	Marshall	1834	Mrs. Allen B. Cummins*	granddaughter-in-law
CUNNINGHAM, Jonathan	Switzerland	1828	William Cunningham	great-grandson
CURTIS, William	Jennings	1827	Curtis W. Russell	great-grandson
CUSHMAN, Isaac	Hancock	1838	J. Russell Cushman	grandson

In 1837 my grandfather, Bartlett Woods, bought this farm from the U.S. government for $1.25 per acre. When grandfather was no longer able to run it my Father bought out rest of family & he ran it until he too was physically unable to. My husband & I have been here over twenty one years.

Mrs. C. A. Clark
Lake County

IHS BASS PHOTO CO. COLLECTION, #267771F3

Rappite wagon from New Harmony, 1946

ANCESTOR	COUNTY	DATE	DESCENDANT	RELATIONSHIP
CUTSINGER, Jacob	Johnson	1823	Thomas A. Cutsinger	grandson
DALLAS, Joe	LaGrange		Dora Dallas*	granddaughter-in-law
DANIELS, Reuben Grant	DeKalb	1839	Pearl (Daniels) Fretz Stewart	granddaughter
			Neil Arthur Waterman	great-grandson
			Ross E. Waterman	great-grandson
			Hazel (Daniels) Wittmer	great-granddaughter
DAUGHERTY, Robert	Orange	1818	Ray A. Daugherty	great-grandson
DAVID, Jacob	Rush	1830	Frank C. Gray	great-grandson
DAVIDSON, John	Hancock	1835	Porter Bolander	great-grandson
DAVIDSON, Joseph	LaGrange	1832	J. L. LaRue	great-grandson
DAVIDSON, Wm. Fleming	Montgomery	1827	Clayton Kessler	great-grandson
DAVIS, Beacham	Jefferson	1834	William Howard Davis	grandson
DAVIS, Evan	Hendricks	1822	D. Wesley Davis	great-grandson
DAVIS, Joseph	Wayne	1823	Mrs. Harmon Dennis	great-granddaughter
DAVIS, Rezin	Shelby	1832	John E. Harper	great-grandson
DAVIS, Robert	Clinton	1836	L. D. Davis	grandson
DAWALT, Henry	Washington	1809/14	Earl W. Dawalt	great-grandson
			Loran B. Payne	great-great-great-grandson
DEAL, Thomas O.	Scott	1846	Lee J. Austin	great-grandson
			Thomas E. Deal	grandson
DEARINGER, David	Rush	1837	Ert Dearinger	grandson

ANCESTOR	COUNTY	DATE	DESCENDANT	RELATIONSHIP
DECKARD, Michael	Monroe	1828	Elmer Deckard	great-grandson
DECKER, John A.	Pike	1836	Charles W. Bradfield	grandson
			Clarence A. Bradfield	grandson
DEER, Lewis	Johnson	1830	Viola Etter	great-granddaughter
DEETER, Jacob	Clay	1841	Frank E. Deeter	great-grandson
DEFOREST, Joseph	Warrick	1818	Sylvester T. DeForest	grandson
DEHART, Adam	Tippecanoe	1824	Elmer Ritchey Waters	great-grandson
DEHAVEN, Isaac	Fayette	1836	Mrs. Elmer A. Edkins	great-granddaughter
			Bessie Moffett*	granddaughter-in-law
DEMAREE, Peter	Johnson	1827	George L. Jeffery	great-grandson
DENNY, John	LaGrange	1835	Madalena Denny	granddaughter
DEVIN, Alexander	Gibson	1813	Anna Hudelson	great-granddaughter
			Arthur Whitsett	great-grandson
DEVORE, Thurett	Johnson	1834	Philander DeVore	grandson
DICE, Michael	Miami	1847	Claude A. Dice	grandson
DICKEY, James	Fayette	1823	Hugh Dickey	grandson
DICKEY, William	Fayette	1821	Florence Elliott	great-great-granddaughter
DICKINSON, Barret	LaGrange	1843	Artimus V. Dickinson	great-grandson
DICKINSON, George	LaGrange	1837	Artimus V. Dickinson*	grandnephew
DICKMEYER, George Henry	Jackson	1839	Harry Dickmeyer	great-grandson
DICKSON, Levi	Tippecanoe	1827	William T. Bull	great-grandson

Originally had 3 fireplaces, one each up & down. . . . The bake oven was on the south west corner and extended about 6 feet beyond the building. . . . The building is in still a very good state of preservation after one hundred and eight years of weathering the elements.

Fred C. LaTourette
Fountain County

IHS DAVID PEAT COLLECTION

Rumely Oil Pull

Joseph Trimble paid the U.S. government $100 for the 160 A[cres]. 26 yrs. later [1846] my father [Elijah McKnight] paid $1260 for the same acreage. My father built an 8 room colonial house in 1861 which has housed his family all thro the years, and still holds up its head proudly.

Epha M. Fishback
Lawrence County

ANCESTOR	COUNTY	DATE	DESCENDANT	RELATIONSHIP
DIEFENBACH, Michael	Dearborn	1841	Harry Diefenbach	great-grandson
			Mary Diefenbach	great-granddaughter
DIEHL, Isaac, Sr.	DeKalb	1840	James G. Diehl	grandson
			Lawrence Diehl	grandson
DILLEY, Joseph	Sullivan	1817	Carrie E. Rhodes	great-granddaughter
DILLMAN, Daniel	Cass	1839	Carl Dillman	grandson
DINIUS, John	Huntington	1846	Boyd S. Dinius	grandson
DITMARS, Henry Stryker	Noble	1843	Dale Schenck	great-grandson
			Homer Schenck	great-grandson
DIXON, Ebenezer	Putnam	1832	Henry Dixon	not given
DIXON, Ebenizer	Montgomery	1832	Henry H. Dixon	great-grandson
DIXSON, Eli	Greene	1825	Mary (Dixson) Sloan	great-granddaughter
DOAK, Joseph W.	Orange	1819	Milton Mavity	great-great-grandson
DOEHRMAN, Conrad, Sr.	Adams	1839	Martha A. (Doehrman) Sonnigsen	great-granddaughter
DOERR, Jacob Peter	Jackson	1838	John W. Heller	grandson
DOLL, Anthony	Ripley	1841	Wilbur Doll	great-grandson
DONABAN, William	Cass	1837	Grace Carroll	great-granddaughter
DOOLEY, Robert	Boone	1838	Bessie F. Neese	great-granddaughter
DORWIN, Calvin T.	Adams	1824	Sherman Kunkel	great-grandson
DOUGLAS, Jesse	Gibson	1835	Hershel Douglas	great-grandson
DOUP, George	Bartholomew	1820	George Doup	great-great-grandson
			Perry Doup	great-great-grandson
DOUTHITT, Thomas M.	Sullivan	1832	Mrs. George E. Douthitt*	great-granddaughter-in-law
			Mrs. Wm. T. Douthitt*	great-granddaughter-in-law
DOW, William	Jefferson	1818	William Dow	grandson
DRAGOO, Peter	Delaware	1830/34	Garland W. Dragoo	grandson
			Claude C. Dragoo Sr.	great-grandson
DRISCOLL, Daniel	Wabash	1834	Clay Driscoll	grandson
DRULEY, Nathan	Wayne	1844	Earl B. Druley	grandson
DUBOIS, Smith	Union	1854	Arthur Dubois	great-grandson
DUFOUR, John David	Switzerland	1803	Olive (Dufour) Trafelet	granddaughter
DUGAN, John	Grant	1837	Adam A. Hilliard	great-grandson
DUGUID, John	Steuben	1844	J. C. McNaughton	grandson
DULING, Solomon	Grant	1837	Virgil B. Duling	grandson
DUNHAM, ——	Hancock	1839	Wm. F. Dunham	grandson
DUNN, Harmon	Grant		Mary (Littler) Browning	granddaughter
DUNN, John	Grant	1838	J. Homer Dunn	great-grandson
			Philip L. Dunn	great-grandson

ANCESTOR	COUNTY	DATE	DESCENDANT	RELATIONSHIP
DUNTEN, Thomas	Allen	1834	L. H. Dunten	not given
DYKES, Henry	Parke	1838	Howard Stark	grandson
			Ray Thomas	grandson
EAGLESFIELD, Thomas	Marion	1834	Derexe (Scudder) Andrews	granddaughter
			Jane Scudder	granddaughter
EASTMAN, Nathaniel	Jennings	1818	Harold Eastman	great-great-grandson
EATON, William	Union		Durward L. Eaton	great-grandson
			James H. Eaton	great-grandson
			Scott V. Eaton	great-grandson
EGLER, Wendel	Dubois	1840	Edward Egler	grandson
EIKENBERRY, Isaac	Carroll	1832	Sarah (Angle) Spitler	great-great-granddaughter
ELDER, Samuel	Vermillion	1832	James P. Elder	great-grandson
ELDER, William M.	Decatur	1822	Clifford Elder	grandson
ELIASON, Joshua	Wayne	1814/31	Gaar G. Eliason Sr.	grandson
ELLIOTT, ——	Posey	1836	Morton T. Elliott	grandson
ELLIOTT, Richard	Harrison	1843	Robert W. Elliott	not given
ELLIS, Mordecai N.	Carroll	1841	Ellis Arthur Hopkins	great-grandson
			James Ellis Arthur Hopkins	great-grandson
			John Ellis Hopkins	great-grandson
			Martha (Ellis) Hopkins	granddaughter
ELPERS, Bernard	Vanderburgh	1847	Katie Stolz	granddaughter
ELY, Henry	Tippecanoe	1825	J. S. Ely	grandson
EMERSON, James	Tippecanoe	1824	Kate Opp	granddaughter
EMERSON, Reuben	Posey	1814	Jesse P. Emerson	grandson
EMERSON, William	Dearborn	1832	Ralph W. Emerson	grandson
EMISON, Thomas	Knox	1802	Samuel M. Emison	great-grandson
EMLEY, John R.	Huntington	1834	Faith (Emley) Rahn	great-granddaughter
EMMONS, John	Fulton	1841	Allene (Emmons) Biddinger	great-granddaughter
EMPSON, John	Jackson	1823	Marion Empson	great-grandson
ENGLE, William	Randolph	1842	Jennie M. (Engle) Staoudt	granddaughter
EVANS, William	Decatur	1834	Glenn R. Evans	grandson
EWBANK, John	Dearborn	1811	Loebo J. Ewbank	great-grandson
EWERY, Sam	Tippecanoe	1837	Ralph C. Fouts	great-grandson
EWING, Patrick	Decatur	1828	Louis Ewing	great-great-grandson
FAIRCHILD, Barton	Jay	1840	Robert Fairchild	grandson
FALL, Christian	Putnam	1820	Frances (Epperson) Winslow	great-great-granddaughter

Large, brick Colonial home built in 1841 is still occupied.

Kathryn Albaugh Kline
Martin County

A cottonwood tree is still standing under which Michael Wilson lived until he got his cabin built in 1834. This is on the 40 acres obtained from the government.

Evan W. Wilson
Madison County

ANCESTOR	COUNTY	DATE	DESCENDANT	RELATIONSHIP
FARDEN, James Nelson	Warren	1847	Millard Farden	grandson
FARMER, Eli P.	Monroe	1830	Anna (Farmer) Wood	granddaughter
FARRINGTON, Wellington	DeKalb	1845	Mary E. Thomson*	granddaughter-in-law
FEE, John	Steuben	1835	Paul Fee	great-grandson
FELKNER, Jacob	Kosciusko	1837	Esther (Felkner) Bates	granddaughter
			Chester M. Felkner	grandson
			Eugene W. Felkner	grandson
			George W. Felkner	grandson
			Lloyd J. Felkner	grandson
FERGERSON, George W.	LaGrange	1836	Mary Jane Fergerson*	daughter-in-law
FERGUSON, John	Floyd		Lizzie (Ferguson) McPheeters	daughter
FERGUSON, Noah	Boone	1831	Bert L. Ferguson	great-grandson
FERGUSON, Thomas	Vigo	1819/31	Emma (Davis) Hunt	great-granddaughter
			Mrs. James E. Mackell	great-granddaughter
FERRELL, James	Hancock	1838	Ward Francis McCarty	great-grandson
FICKLE, William	Clinton	1832	McClellan Fickle	grandson
FIDLER, Jerry	Vigo	1837	John R. Fidler	great-grandson
			William W. Fidler	great-grandson
			Franklin O. Fidler	great-grandson
FIFER, Joseph	Marshall	1845	Wm. E. Fifer	grandson
FINK, Andrew	Hancock	1835	Marie (Fink) Bardomer	great-granddaughter

Rural scene near Crawfordsville

IHS BASS PHOTO CO. COLLECTION, #218696F

ANCESTOR	COUNTY	DATE	DESCENDANT	RELATIONSHIP
FINLEY, James M.	Clay	1845	Crystal Finley	granddaughter
			Lois Finley	granddaughter
			Pearl Finley	granddaughter
FISHER, Jacob	Johnson	1825	Erie R. Fisher	grandson
FISHER, John	Marshall	1845	G. T. Bigley*	grandnephew
FITCH, Nathaniel	DeKalb	1835	Gladys Fitch	granddaughter
			Marie Houser	granddaughter
			Marguerette Wert	granddaughter
FITZER, Joseph	Cass	1832	Harry J. Fitzer	grandson
FLINN, William	Lawrence	1830	Ella B. Callaway	great-granddaughter
FLOOD, Benjamin	La Porte		Mary M. Dolman	granddaughter
FOGLESONG, Christian	Johnson	1830	Velma F. Thomas	great-granddaughter
FOLAND, George	Hamilton	1825	Perry H. McClintock	great-grandson
FOLKENING, Charles	Marion	1843	Urban Folkening	great-grandson
FORCE, Whitfield	Martin	1835	Dorothy Baker	great-granddaughter
			Charles Force	great-grandson
			Emma Force	great-granddaughter
			Hazel Force	great-granddaughter
			Catherine (Force) Lang	great-granddaughter
			Gertrude Weathers	great-granddaughter
FORRESTER, James	La Porte	1840	James Forrester	grandson
			Martha Forrester	granddaughter
FOSTER, Benjamin	Fountain	1825	Lee Foster	great-great-grandson
			Paul Foster	great-great-great-grandson

Thomas Miler came into possession of the Dotson Grant and later my father Chas. T. Myler purchased same land from his father, later it came into my possession and in 1928, my husband & I sold it to Edward Faulkenburg (present owner) whose wife is a Great Granddaughter of Thomas Miler.

Trunie M. Neal
Crawford County

IHS INDIANA EXTENSION HOMEMAKERS COLLECTION

There is an undying spring said to be an old Indian Waterhole. It was walled up with a hollow log standing on end over the main source of water. It was about five feet deep. We have Indian arrow-heads and other Indian relics which have been found on our farm.

Mrs. John LaMont
Franklin County

ANCESTOR	COUNTY	DATE	DESCENDANT	RELATIONSHIP
FOSTER, John	Tippecanoe	1828	Mary Bryan	great-granddaughter
			Mrs. O. B. Leonard	great-granddaughter
			Mrs. George E. Shelby	great-granddaughter
FOSTER, Robert	Montgomery	1831	E. Frank Wilkinson	grandson
FOUST, Jonathan	Huntington	1839	Emma A. (Foust) Irwin	granddaughter
FOUTS, George	Carroll	1831	G. Earl Fouts	grandson
FOUTS, Michael	Miami	1837	Mrs. N. B. Brower	granddaughter
FOUTS, William	Wayne	1819	Martha C. Carey	great-granddaughter
FOX, Joseph	Knox	1842	Lee Fox	grandson
FRAILEY, Christian, Sr.	Elkhart		Ella (Frailey) Danielson	granddaughter
FRAKES, Asa	Vigo	1841	Etta J. (Jewell) Logan	great-great-granddaughter
FRANCIS, James	Montgomery	1833	Annie V. Meharry*	grandniece
			Ira G. Meharry*	grandnephew
FRAZEE, Ephriam	Rush	1820	Dorothy Douglas	granddaughter
			Ed Frazee	grandson
FRAZER, Benjamin Franklin	Tippecanoe	1830	Charles H. Frazer	grandson
FREDRICKSON, Henry	La Porte	1836	Ella G. Andrew	great-granddaughter
FREELAND, William	Greene	1846	Florence E. Hendricks	granddaughter
FREEMAN, Isaac	Delaware	1837	Ernest Freeman	great-grandson
FREESE, Israel	Marshall	1847	Mrs. Howard Waltz	granddaughter
FRENCH, Philip B.	Gibson	1838	Jesse B. French Sr.	grandson
FRIE, John Frederick	Jackson	1837	Oscar J. Frey	great-grandson
FRIEND, Henry	DeKalb	1834	Arthur M. Friend*	great-grandnephew
FRY, John, Sr.	Clark	1826	John E. Lentz	grandson
FRYBARGER, George	Fayette	1829	Clifford Brattain	great-grandson
FULGHUM, Jesse	Wayne	1826	Thelma (Overman) Moody	not given
FULTON, Samuel	Huntington	1835	Howard Fulton	grandson
FUNK, Henry	Harrison	1832	Clara Funk	granddaughter
FURNAS, John	Hendricks	1836	Albert L. Copeland	great-grandson
GAGEBY, John	Decatur	1821	Wood H. Gageby	great-grandson
GALLOWAY, James	Orange	1824/32	Emma Ruth (Galloway) Bell	granddaughter
			Dora (Wolfe) Coulter	granddaughter
			Laura B. Galloway*	granddaughter-in-law
GAMBLE, Thomas	Wabash	1848	Orville Gamble	grandson
GARDNER, Philip	Putnam	1826	Fletcher Goff	grandson
GARMAN, Enoch	Allen	1847	Eli H. Garman	son
GARRIOTT, Simeon M.	Washington	1818	Homer A. Garriott	grandson

Plowing with mules

IHS BASS PHOTO CO. COLLECTION, F27-6

ANCESTOR	COUNTY	DATE	DESCENDANT	RELATIONSHIP
GARRISON, John	Carroll	1834	Henry Garrison	grandson
GASKILL, William	Clinton	1834	A. Ray Gaskill	grandson
GATES, Richard	Union	1825	Catherine (Pauley) Masters	great-great-granddaughter
GAY, William	White		Edward C. Gay	great-grandson
GEHLHAUSEN, Francis G.	Dubois	1846	Leo G. Gehlhausen	grandson
GEIGER, John	Jay	1847	Lola (Addington) Bourne	granddaughter
GEORGE, Milton	Jefferson	1834	Russell George	great-grandson
GERARD, Andrew R.	Marshall	1845	Charles C. Gerard	son
GERBER, Abraham	Miami	1848	Ezra F. Feller	grandson
GEYER, John	Elkhart	1843	Verda Slabaugh	great-granddaughter
GIBBONS, John	Marshall	1846	Mrs. F. M. Gibbons*	daughter-in-law
GIBBS, Hugh	Hancock	1837	William W. Martindale	great-grandson
GILBERT, Owen G.	LaGrange	1849	Owen G. Gilbert	grandson
GILLESPIE, Thomas	Scott	1839	John D. Gillespie	grandson
GILTNER, Jacob	Clark	ca. 1790	Wm. A. Giltner	grandson
GIRTON, Steven	Clay	1838	Elias W. Girton	grandson
GLADDEN, Joseph	White	1845	Catharine (Hutton) Black	great-granddaughter
GLASCOCK, Francis J.	Fountain	1845	Samuel J. Glascock	grandson
GLASCOCK, Thomas	Hancock	1838	Albert J. Glascock	grandson
GLENDENNING, Harry W.	Boone	1852	Frank Glendenning*	grandnephew
GLIME, John	La Porte	1846	George Hager	grandson
GLOYD, George	Allen	1837	Estella Gloyd	granddaughter

They landed at Troy, Indiana about mid-summer of 1837 and there they hired a man with an ox team and wagon. . . . When they were about twenty-four miles north of Troy, which is now in the vicinity of Bretzville . . . a German by the name of Melchior Jerger . . . advised them to settle in his neighborhood. . . . They built a log cabin thereon and then became settled.

Jacob H. Bretz
Dubois County

Mill machinery, brought in from Kentucky, was erected here by Thomas Emison, and was declared a public utility by the Territorial Court in 1807. The fort which protected it [the mill], was nicknamed "Fort Petticoat" because [the fort's store was] run by "Maria and the girls."

Emily Adams Emison
Knox County

ANCESTOR	COUNTY	DATE	DESCENDANT	RELATIONSHIP
GODDARD, Joseph	Decatur	1824	Fred Goddard	great-grandson
GOODLANDER, Jacob	Wabash	1837	Bessie F. Campbell	not given
			Bernice (Campbell) Sutherlin	not given
GOODMAN, Micajah Torrey	Vigo	1819	Mrs. Fred M. Goodman*	great-granddaughter-in-law
GOODWINE, James	Warren	1829	John G. Crone	great-great-grandson
GORDON, John	Rush	1838	Effie (Offutt) Bagley	great-great-granddaughter
			Clark Offutt	great-great-grandson
GORDON, Uriah	Rush	1847	Effie (Offutt) Bagley	great-granddaughter
			Clark Offutt	great-grandson
GOSHERT, John	Kosciusko		LeRoy Wilson Goshert	grandson
GOSS, Joseph	Jackson	1837	John F. Goss	great-grandson
GOSS, Joseph	Jackson	1843	Chester A. Goss	great-grandson
GOSSMAN, Henry	Jackson	1848	Ralph Gossman	grandson
GOTTMAN, Andrew	Vanderburgh	1847	Charles A. Gottman	grandson
GRABLE, Samuel	Cass	1845	Mrs. John J. Brackett	great-granddaughter
			Mrs. John I. Mason	great-granddaughter
			Mrs. Ford Smith	great-granddaughter
GRAHAM, Elizabeth	Madison	1836	Edna Jane Graham	great-granddaughter
			Lois M. Graham	great-granddaughter
			Omer J. Sears	great-grandson
			Arna Smultz	great-granddaughter
GRAINGER, Ira P.	Vanderburgh	1832	Richard T. Legler	great-grandson
GRANTHAM, John L.	Carroll	1834	Wilber L. Grantham	great-grandson

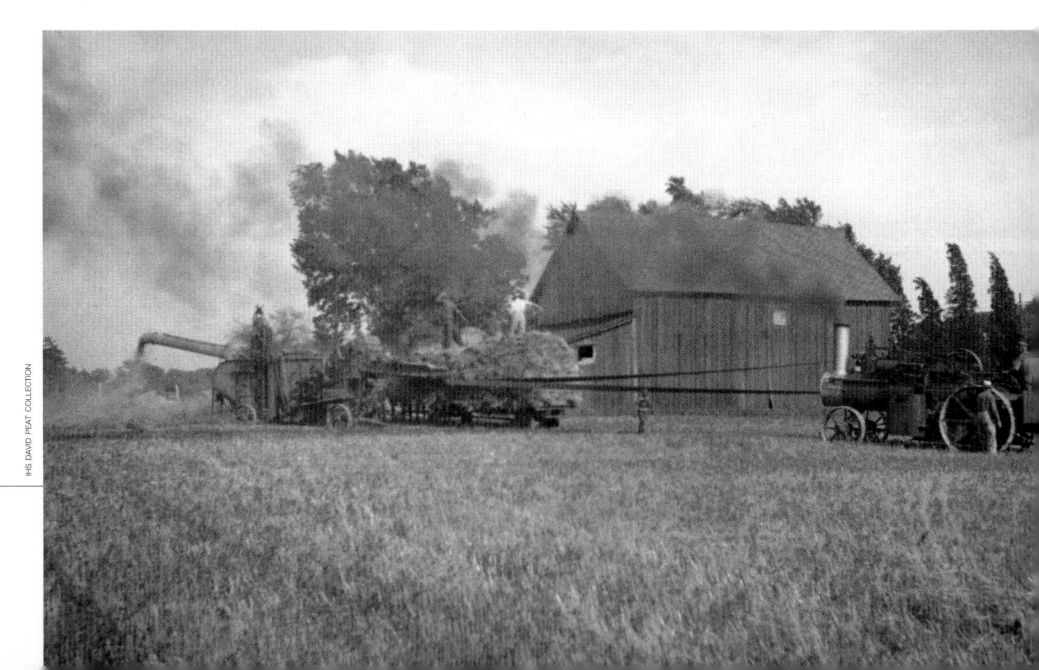

IHS DAVID PEAT COLLECTION

Threshing scene, Don Schewenk Farm

IHS INDIANA EXTENSION HOMEMAKERS COLLECTION

ANCESTOR	COUNTY	DATE	DESCENDANT	RELATIONSHIP
GRAY, Ebenezer	Switzerland	1826	Uly G. Gray	grandson
GRAY, James	Rush	ca. 1816	J. L. Norris	great-grandson
GRAY, Jesse	Jay	1836	James H. Williams	great-grandson
GRAYHAM, Hezekiah	Rush	1846	Buell Graham	grandson
GREEN, Jacob	Shelby	1840	Arthur B. Barnard	grandson
GREENWOOD, Henry Bailey	Morgan		Carl E. Brown	great-great-grandson
			Ralph O. Brown	great-great-grandson
			William Hugh Brown	great-great-grandson
			Marian (Brown) Small	great-great-granddaughter
			Verna (Brown) Vandeventer	great-great-granddaughter
GREGORY, James	Montgomery	1831	Mills Bratton	great-great-grandson
GRIFFIN, Arstarchus	La Porte	1840	Capitola Griffin	daughter
GRIMES, George	Wayne	1838	Ruth A. Campbell	granddaughter
			Nellie L. Davenport	granddaughter
GRISELL, Samuel	Jay	1836	Lewis W. Hoover	great-grandson
GROSECLOSE, David	Johnson		Oletha G. Deer	great-granddaughter
GROSS, Martin	Madison	1837	Frank B. White	grandson
GRUBBS, ——	Dearborn		Theodore Grubbs	grandson
GUFFIN, Andrew	Rush	1821	Florence (Guffin) Wilson	great-granddaughter
			Dora (Guffin) Wood	great-granddaughter
GULLEFER, Aaron	Marion	1825	Harry R. Gullefer	great-grandson
GUSTIN, Samuel B.	Madison	1836	Arthur J. Gustin*	great-grandnephew

They were good citizens. . . . The Indians helped Reuben Long clear off the forest and build his first buildings. There has been three house[s] on same spot. The first house destroyed by an Indian attack. They later made friends and help[ed] him clear his land.

Bernice Long
Whitley County

Barges were built on these farms, loaded with grain and pork making the trip to New Orleans to sell the produce. A brick house is still being used which is located on one of these farms. This house was built in 1859 and the bricks for this house were made from clay on the farm and fired there.

Mary Hendricks Kennedy
Morgan County

IHS INDIANA EXTENSION HOMEMAKERS COLLECTION

Threshing machine

ANCESTOR	COUNTY	DATE	DESCENDANT	RELATIONSHIP
GUTHRIE, ——	Marion	ca. 1830	Annie M. Simmons*	great-granddaughter-in law
GUY, Robert	Madison	1832	Myrl Guy New	great-granddaughter
GUZG, John	Rush	1820	John Jones	grandson
GWALTNEY, John	Spencer	1839	Alice Metz	granddaughter
GWARTNEY, William	Harrison	1825	William Eckart	great-great-grandson
GWINN, James	Madison	1835	Ernest L. Gwinn	great-grandson
HACKLEMAN, Richard	Hancock	1845	Alice (Hackleman) Scott	granddaughter
HADSELL, James	DeKalb	1836	Eva Shull*	granddaughter-in-law
HAEHL, George Michael	Shelby	1834	Kurt C. Kuhn	grandson
HAINES, James	Grant	1843	Willis Haines	grandson
HALBERT, Joel	Martin	1815	Joseph Gaddis	great-grandson
HALL, Franklin	Hamilton	1834	Elizabeth Asenath Hall	daughter
HALL, George	La Porte	1845	Shirley M. Kramer	granddaughter
HALL, Isaac	Allen	1837	George William Harter	great-grandson
HALL, Samuel D.	Kosciusko	1837	William Roberts Hall	grandson
HAMILL, Robert	Boone	1835	Burchell Hamill	great-grandson
HAMILTON, Cyrus	Decatur	1821	Tom M. Hamilton	grandson
HAMILTON, Hugh	Spencer	1846	Mrs. Joe L. Mackey	not given
HAMILTON, James H.	Morgan	1832	Ira W. Hamilton	grandson
HAMILTON, John	Clinton	1844	Laura M. Smith	great-granddaughter
HAMILTON, Robert	LaGrange	1837	Burritt Hamilton	grandson

ANCESTOR	COUNTY	DATE	DESCENDANT	RELATIONSHIP
HAMMAN, Thomas	DeKalb	1847	Robert J. Hamman	grandson
HANDLEY, James	Fayette	1821/31	Lourena B. (Handley) Elliott	granddaughter
			Mrs. Quincy Elliott	granddaughter
HANSON, Amos O.	Fayette	1836	B. F. Claypool	great-grandson
HANSON, Benjamin Hollis	Fayette	1814	B. Hollis Hanson	great-grandson
HARDING, George	Steuben	1834	George F. Harding Sr.	grandson
HARLAN, Samuel	Fayette	1813	Jesse W. Harlan	great-grandson
			Samuel J. Harlan	great-great-grandson
HARLAN, Stephen	Hancock	1834	Mrs. Rosa G. Harlan°	granddaughter-in-law
HARNESS, George W.	Carroll	1829	Elsworth Harness	grandson
HARRIS, Thomas	Grant	1834	Ora Eilar	granddaughter
HARRIS, Thomas	Owen	1835	William S. Harris	great-grandson
HARRISON, Benjamin	Sullivan	1845	Tamar De Hart	granddaughter
HARRISON, Robert	Montgomery	1825	Henry A. Harrison	great-grandson
HARSHBERGER, Jacob	Montgomery	1832	Alice Anderson	great-granddaughter
			Anna Anderson	great-granddaughter
			Harry Anderson	great-grandson
			Paul Anderson	great-grandson
			Mrs. William Lee	great-granddaughter
HARVEY, Absalom	Henry		Ralph Harvey	grandson
HARVEY, Andrew	La Porte	1833	Hugh K. Hood	grandson
HARVEY, Isaac	Henry	1847	Alice Bernadina Harvey°	daughter-in-law
HARVEY, John	Wayne	1812	Harry H. Harvey	great-grandson
HARVEY, Robert	Tippecanoe	1830	Ada (Brown) Dailey°	grandniece
HARVEY, Robert	Wells	1831	Henry H. Harvey	grandson
HAUSER, George	Owen	1826	William H. Hauser	great-grandson
HAUSMAN, Phillip	Posey	1845	Louisa Hausman°	daughter-in-law
HAWKINS, John	Huntington	1848	Hazel (Hawkins) Groff	granddaughter
HAWKINS, Samuel	Grant	1848	Oscar T. Hawkins	grandson
HAYES, Richard	Montgomery	1837	Robert D. Hayes	great-grandson
HAYNES, Samuel S.	DeKalb	1843	Dora E. (Haynes) Sliger	great-granddaughter
HAYWORTH, Dillon	Rush	1833	Fred B. White	great-grandson
HAZELWOOD, Polly	Hendricks	1832	Mrs. D. S. Hazelwood°	granddaughter-in-law
HEART, Aaron	Daviess		William Hart	grandson
HEASTON, David	Randolph	1841	Lesta F. Curry	great-great-granddaughter
HEATON, Joseph	Rush	1843	Robert Heaton	grandson
HEATON, Thomas	Rush		Lottie Stark	granddaughter
HEATON, William S.	Owen	1845	Chester A. Heaton	grandson

A round barn 80' in diameter with immense silo in the centre, and drive in hay mow. Frame, hewed oak cut from the farm, ground floor was concreted, two extensive orchards. At one time there were 5 homes on the farm.

William H. Charles
Orange County

When Amos Black took possession [in 1846], sap troughs hewn from block of section of trees, by Indians, were found. Deer were in the woods.

Bess Black Benton
Noble County

ANCESTOR	COUNTY	DATE	DESCENDANT	RELATIONSHIP
HEAVILON, Taylor	Clinton	1829	Jessie A. Heavilon°	daughter-in-law
HEDDEN, George W.	Harrison	1843	William H. Hedden	grandson
HEDRICK, Abraham	Grant	1834	Grace (Kennedy) Lewis	granddaughter
HEILERS, Anton	Dubois	1840	Benno Heilers	great-grandson
HELM, Jefferson	Rush	1837	Fred P. Cutter	grandson
HELMS, Isaac	Hamilton	1834	Grace Bergner°	great-granddaughter-in-law
HENDRICKS, Henry	Vanderburgh	1840	Henry Hendricks	grandson
			John Hendricks	grandson
			Fred Opperman	grandson
HENDRICKS, Thomas	Morgan	1832	Mary (Hendricks) Kennedy	great-granddaughter
HENDRICKS, Thompson	Morgan	1847	Mary (Hendricks) Kennedy	granddaughter
HENDRICKSON, Jacob	Fulton	1841	Oren M. Hendrickson	grandson
HENDRIX, George B.	Carroll	1846	Mrs. Warren Knapp	granddaughter
HENLEY, Joseph	Rush	1822	Jesse Henley	grandson
HENLEY, Micajah	Wayne	1818	Robert M. Henley	grandson
HENLEY, Thomas	Rush	1829	Clyde C. Henley Sr.	grandson
HENRICKS, Jobst	Vanderburgh	1840	Mrs. Elder L. Miller Sr.	great-granddaughter
HENRY, Samuel	Hancock	1837	Gerry Henry	great-grandson
			Samuel J. Henry	grandson
HERLITZ, Lewis	Lake	1840	Louis F. Herlitz	grandson
HERR, Christian	Henry	1839	John R. Herr	grandson
HERTSCHEL, Christian	Rush	1845	Warren Carmony	great-grandson
			Wayne Carmony	great-grandson
			Clarine Webster	great-granddaughter
HIATT, Amos	Randolph	1837	Orpha M. Barnes	granddaughter
HIATT, Richard	Randolph	1845	Nellie (Diehl) Higgins	granddaughter
HICK, Peter	Shelby	1821	Bryan Heck	great-grandson
HICKS, George W.	Vermillion	1822	Daniel A. Hicks	great-grandson
HIESTAND, Manuel	Boone	1837	Charles W. Routh	grandson
HIGMAN, Robert	Tippecanoe	1832	Mrs. William E. McCoy	great-granddaughter
HILL, Benjamin	Wayne	1812	Mrs. —— Hill°	granddaughter-in-law
HILL, John	Rush	1828	Mary B. Bentley	great-granddaughter
HILL, William	Rush	1833	Ruth E. White	great-granddaughter
HINDSLEY, James	Clinton	1848	Ethel (Thatcher) Burnau	granddaughter
HINES, Olney	Warrick	1825	Paul P. Hines	great-great-grandson
HINSHAW, John	Randolph	1839	Mildred Inez (Hinshaw) Brown	granddaughter
HIZER, John	Fulton	1836	Mabel Wills	great-great-granddaughter
HOADLEY, Abram, Sr.	Hendricks		Virgil Hoadley	grandson

ANCESTOR	COUNTY	DATE	DESCENDANT	RELATIONSHIP
HOBAN, John	Whitley	1839	Mary Hoban	granddaughter
HOBBS, Isaac	Scott	1820	Homer D. Elliott	great-grandson
HOCHSTETLER, William	Miami	1848	P. William Hostetler	grandson
HODGES, Thomas	Morgan	1832	Rachel C. Hodges	great-granddaughter
HODSON, Uri	Hamilton	1835	Ernest Hodson	grandson
HOFFMAN, Jacob	Vigo		Oliver Hoffman	grandson
HOFFMAN, John, Jr.	Vigo	1835	Lillian Moore°	grandniece
HOGUE, Joseph	Knox	1841	Lewis Z. Hogue	grandson
HOKER, Frederick	Ripley	1825	Harry Westmeyer	grandson
HOLADAY, Samuel	Orange	1846	Annie Holaday°	granddaughter-in-law
			Evaline Holaday°	granddaughter-in-law
HOLAM, Jacob	Marshall	1845	Olen York	great-grandson
HOLLCRAFT, Abraham	Clinton	1834	Charles B. McClamrock	great-grandson
HOLLINGSWORTH, John	Knox	1848	Jeanne Coan	great-granddaughter
HOLLMANN, John F.	Allen	1835	Paul Hollmann	grandson
HOLLOWELL, William	Randolph	1847	Birch Hollowell	grandson
HOLMES, Samuel	Carroll	1830	Charles L. Wise	great-great-grandson
HOOVER, Andrew	Wayne	1806	Grace (Bulla) Engelbrecht	great-great-granddaughter
HOOVER, David	Wayne	1806/12	David Ralph Hoover	great-grandson
			David S. Hoover	grandson
			Mrs. Larkin Hoover°	granddaughter-in-law
			William Bryan Hoover	great-grandson
			Marie (Hoover) Martindale	great-granddaughter
			Irena (Hoover) Williams	great-granddaughter
HOOVER, George	Henry	1836	Mrs. J. J. Hoover°	granddaughter-in-law
HOOVER, Henry	Fulton	1835	Ford Johnson	great-great-grandson
HOOVER, Martin	Randolph	1852	Mrs. John W. Botkin	granddaughter
HOPPENJANS, Gerhard	Dubois	1842	Bernadina Hoppenjans°	daughter-in-law
HORNER, Jacob	Washington	1816	John Eli Horner	grandson
HORRALL, Thomas	Daviess	1839	Edith (Horrall) Bingham	granddaughter
HOUGHTON, Aaron	Martin	1835	Houghton S. Albaugh	grandson
			Kathryn (Albaugh) Kline	granddaughter
HOUGHTON, William	Daviess	1818	Kenner K. Dilley	great-granddaughter
HOUSE, John	Shelby	1829	Edwin A. Pritchard Sr.	great-grandson
HOUTS, George	Daviess	1815	Mrs. Dale Barber	not given
HOUTZ, Henry	Wells	1836	Edward Houtz	grandson

Two log houses preceded the present, colonial frame house, which is in very fine condition despite the fact that it is 107 years old. There is still a considerable quantity of virgin timber on the land. The farm is still producing well & is being farmed.

Geo. A. Obery
Ripley County

On December 29, 1849, with his brother John and ten others, [William W. Goodman] organized the West Vigo Congregation Society and at once inaugurated the preliminaries for erecting a house to worship in, which resulted in the dedication of West Vigo Church—May 1, 1853.

Mrs. Fred Goodman
Vigo County

ANCESTOR	COUNTY	DATE	DESCENDANT	RELATIONSHIP
HUBBARD, Ira G.	Elkhart	1843	Annette (Fieldhouse) Causey	granddaughter
			Charles H. Fieldhouse	grandson
			Carrie E. (Fieldhouse) Mathias	granddaughter
HUBBELL, John, Jr.	Fayette	1846	Clinton Hubbell	grandson
HUBER, Godfrey	Franklin	1820	Harvey F. Huber	grandson
HUDKINS, Richard	Fulton	1837	Eva Hudkins*	daughter-in-law
HUEBNER, Lawrence Martin	Posey	1843	Mrs. B. E. Cooper*	niece
HUFF, John	Grant	1837	Adam A. Hilliard*	great-grandson-in-law
HUFFERT, Abraham	Carroll	1831	Vernon Hufford	grandson
HUFFMAN, Aaron	Johnson	1830	Mark Huffman	grandson
HUFFMAN, Henry	Wells	1837	Ardella M. (Huffman) Goodmiller	granddaughter
HUFNAGLE, John George	Vanderburgh	1837	August Hufnagel	grandson
			Henry Hufnagel	grandson
HUGHES, John	Decatur	1848	Dan K. Hughes	great-grandson
			John N. Hughes	great-grandson
			Mary E. McCullough	great-granddaughter
			Florence O. Price	great-granddaughter
			Herminone J. Radcliff	great-granddaughter
HUGHES, Margaret	Scott	1838	Galen Middleton	great-great-grandson
			Cynthia (Coats) Palmer	great-great-granddaughter

IHS INDIANA EXTENSION HOMEMAKERS COLLECTION

ANCESTOR	COUNTY	DATE	DESCENDANT	RELATIONSHIP
HULLER, Jothan	Owen		Silas Huller	grandson
HUMPHREY, John G.	Decatur	1837	Jesse G. Humphrey	great-grandson
HUNGATE, John	Shelby	1833	Roy E. Hungate	great-grandson
HUNT, Libni	Rush	1839	Elmer Hunt	grandson
HUNT, Timothy	Wayne	1815	Russell W. Hunt	great-great-grandson
			William A. Hunt	great-grandson
HUNTER, Matthew Reed	Marion	1845	Adah Hunter	granddaughter
			Lee Hunter	grandson
HUNTZINGER, John	Madison	1851	Hazel Troutman	great-great-granddaughter
HUNTZINGER, Noah	Madison	1859	Mattie H. Hersberger	granddaughter
HURST, George	Putnam	1828	Maude (Hurst) McNary	granddaughter
HUSSEY, Jonathan	Fayette	1836	Florence Hussey	granddaughter
			John L. Hussey	grandson
			Sarah Hussey	granddaughter
HUSTON, John	Madison	1835	Huston Frazier	great-grandson
			Roxana Frazier	great-granddaughter
HUTCHESON, Randolph	Putnam	1834	Philip B. Hutcheson	grandson
HUTCHINSON, Volney Powers	Vigo	1845	Murrel Hutchinson	grandson
HUTTON, Cornelius	Jasper	1851	Grace Tullis	granddaughter
HYDE, Amasa	Switzerland		Albert B. Hyde	grandson
HYRE, Solomon	Cass	1831	Gretna (Robertson) Bowman	great-granddaughter

Four generations of the Wright family now live on this land. . . . It is also interesting to know that the grandson Amos 2nd and his son Shelby Markland 2nd now live in the house built before the Civil War by Isaac Wright son of Isaac B. the original owner.

Jesse Markland Brown
Spencer County

IHS INDIANA EXTENSION HOMEMAKERS COLLECTION

James Christy . . . entered this farm in 1826 when my father was 5 yrs. of age. My father John B. Christy died in 1915 at age of 94 yrs. . . . [I] still live on this farm which is 121 yrs old. I am still on farm & doing my work. I believe this farm to be the oldest in Put[nam] Co.

H. C. Christy
Putnam County

ANCESTOR	COUNTY	DATE	DESCENDANT	RELATIONSHIP
ICEBERG, Frederick	Ripley	1843	George Iceberg	grandson
IMEL, Samuel	Montgomery	1834	Josephine Imel	granddaughter
INNIS, Nathaniel	Rush	1824	John J. Frazier	great-great-grandson
IRELAND, Thomas	Carroll	1835	Wilson B. Ireland	grandson
IRWIN, Elim	Rush	1830	Ortis Werking	grandson
JACKMAN, Hugh	Marshall	1844	Faye Shively	granddaughter
			Jess Shively	grandson
JACKSON, Enoch W., Sr.	Dearborn	1835	P. E. Jackson	grandson
JACKSON, Jeremiah	Clark	1837	Stella (Jackson) Daugherty	granddaughter
JACKSON, Reuben C.	Dearborn	1838	Floyd S. Jackson	grandson
JACOBY, Peter	Clinton	1832	L. S. Jacoby	grandson
JACQUESS, Jonathan	Posey	1814	Katharine S. Jacquess	great-great-granddaughter
JAMES, George W.	LaGrange	1845	John F. James	grandson
JELLEFF, Cyrus	Jay	1838	Clarence O. Jelleff	great-grandson
JESSUP, Thomas	Rush	1839	John B. Jessup	not given
			Mrs. Cecil Pike	not given
JINKENS, Daniel	Madison	1846	Wade H. Bevelhimer	grandson
JOHNSON, ——	Randolph	1829	Cyrus Johnson	great-grandson
JOHNSON, Daniel	Sullivan	1818	Carrie E. Rhodes	great-great-granddaughter
JOHNSON, Edward	Cass	1837	Edward T. Johnson	grandson
JOHNSON, Elias	Posey	1828	Arthur L. Johnson	not given
JOHNSON, Jesse J.	Lawrence	1816	Jesse J. McKnight	grandson
JOHNSON, Joseph, Jr.	Jackson	1834	Florence (Johnson) Hall	great-granddaughter
JOHNSON, Samuel	Knox	1829	Samuel C. Johnson	great-great-grandson
JOHNSON, William H.	Vigo	1834	W. H. Johnson	grandson
JOHNSTON, Henry	Warren		C. Henry Baum Jr.	great-grandson
JONES, Aaron	Vermillion	1835	Willie A. Jones	great-grandson
JONES, Ben	LaGrange		Jennie Troyer	granddaughter
JONES, James H.	Delaware	1841	Mornay Jones	grandson
JONES, John	Greene	1818	Charles J. Jones	great-grandson
			J. J. Jones	great-grandson
JONES, John B.	Spencer	1846	P. W. Jones*	grandnephew
JONES, Reuben	Rush	1831	Maude Jones	great-granddaughter
JONES, Samuel	Huntington	1835	Emma N. Rauch*	great-granddaughter-in-law
JONES, Thomas	Jefferson	1825	Clark M. Jones	great-grandson
JONES, Thomas A.	Allen	1838	Gladys (Jones) Webster	granddaughter
JONES, William C.	Fayette	1814	Nellie V. Stelle	great-great-granddaughter

IHS INDIANA EXTENSION HOMEMAKERS COLLECTION

ANCESTOR	COUNTY	DATE	DESCENDANT	RELATIONSHIP
JONES, William C.	Hancock		Hazel (Lain) Leslie	great-granddaughter
JORDAN, William	Delaware	1845	Francis W. Jordan	great-grandson
JOSLIN, Harrison	Vigo	1835	Iva (Joslin) Trimmer	granddaughter
JUDAY, Jacob	Elkhart	1836	Mayane Juday°	granddaughter-in-law
JUSTICE, James	Cass	1834	Jerome Justice	son
KEATON, Benjamin John	Shelby	1834	Lottie (Hasket) May	granddaughter
KEELER, Parker	Marion	1824	Jot E. Foltz	grandson
KEESLEY, Daniel	Delaware	1832	Martha Shull	great-granddaughter
KEISLING, Will	Rush	1831	Russell Kinnett	great-great-grandson
KEITH, Adam	La Porte	1837	Harry Shoemaker	great-grandson
KELLER, George	Shelby	1840	Lora B. McClain	great-granddaughter
KELLER, Phillip	Clay	1839	George E. Keller	grandson
KELLY, John B.	Warrick	1811	Jess B. Harpole	great-grandson
KENNEY, Jonathan W.	Tippecanoe	1840	Pierre I. Kenney	grandson
KERCHEVAL, James	Hamilton	1838	Emma K. Johnson	granddaughter
KERR, Alexander	Fayette	1821	Chester Kerr	grandson
KERSEY, James C.	Hendricks	1832	Theodore Kersey	great-grandson
			Virgil Kersey	great-grandson
KERSTIENS, Gerhard H.	Dubois	1838	George B. Kerstiens	grandson
KETCHAM, Foster	Warrick	1847	Sanford L. Ketcham	not given
KIGER, Daniel	Hendricks	1832	Fred Worrell	great-grandson

In 1833 at the age of twenty-six, [John Pickett Noble] came from New York state to Laporte Co., on horseback, where he worked for $10 per month. . . . The first school in Jackson Township was taught by Mary Ann Noble in her home. Mr. Noble was an energetic untiring worker clearing his land by day, making ax handles ox bows and yokes of nights and stormy days. He was friendly to the pioneer covered wagon Westbound travelers, helping to fit their ox teams with bows and yokes.

Col. C. C. Gregg
Porter County

John Burton built a church on his land in 1831 and rode horse back to Baltimore to secure a minister. The first Methodist Conference in Indiana was held in this church in 1831. Known as Liberty Methodist Protestant.

Mr. and Mrs. Dewey D. Douglas
Monroe County

ANCESTOR	COUNTY	DATE	DESCENDANT	RELATIONSHIP
KINCAID, Frederick	Boone	1847	N. N. Kincaid	grandson
KING, John	Wayne	1818	Kermit King	great-grandson
KING, William	Rush	1842	Marie (King) Booth	granddaughter
KINGSBURY, Hiram	Lake	1844	Mrs. A. Logan Steele	great-granddaughter
KINGSLEY, Simeon	Harrison	1820	Rachel Gertrude Kingsley°	great-granddaughter-in-law
KINNAMAN, Walter	Madison	1831	Ray Smith	great-great-grandson
KINNEAR, Michael	Jefferson	1814	Mrs. Harry Underwood	great-granddaughter
KINSEY, David	Whitley	1836	Carl Kinsey°	grandnephew
			Harry Kinsey°	great-grandnephew
			Hubert Kinsey°	great-grandnephew
KINTNER, Jacob Lamb	Harrison	1829	Julia (Kintner) Withers	granddaughter
KIRBY, Benjamin	Decatur	1834	Mrs. Carl Swift	great-granddaughter
			Fred Woolverton	great-grandson
			Victoria Woolverton	great-granddaughter
KIRKPATRICK, David	Rush	1826	David T. Kirkpatrick	grandson
KIRLIN, William	Wayne	1832	John Kirlin	great-grandson
KIRLING, Ephraim	Henry	1837	Mrs. Will Burrow°	grandniece-in-law
			Mrs. Aron Peckinpaught°	grandniece-in-law
KITCHEL, Samuel	Union	1815	Bernard Kitchel	great-great-grandson
KLEINSCHMIDT, Conrad, Sr.	Posey	1835	Conrad Kleinschmidt Jr.	son
KLINE, Jacob, Sr.	LaGrange	1852	Jacob Kline Jr.	son
KLINE, James, Sr.	LaGrange	1852	James Kline Jr.	son
KLINGENSMITH, Joseph	Marion	1829	Laura A. (Klingensmith) Arrman	granddaughter
KNIGHT, Levi	Shelby	1835	Eleazar Knight Amsden	grandson
KNIPE, John	Owen	1845	Gladys Haltom°	step-granddaughter
KNIPPI, George	Clay	1839	Leona (Strauch) Keiser	great-granddaughter
KNOOP, Michael	Huntington	1837	M. K. Bussard	grandson
			Sarah Bussard	granddaughter
KNOWLES, Eli	Gibson	1826	Forman E. Knowles	grandson
KOERNER, John G.	Union	1832	Fred H. Brookbank	great-grandson
KOONS, Joseph	Henry	1835	Benjamin E. Koons	grandson
KRUMMEL, Jacob	Ripley	1838	Charles Krummel	son
KRUPP, Joseph	St. Joseph		Merle Krupp	grandson
KUNKEL, George	Montgomery	1832	Iva G. Davidson	granddaughter
KUNKLER, Fred	Dubois	1839	Albert Kunkler Sr.	not given
LAFUSE, John	Union	1838	Lester LaFuse	grandson
LAGRANGE, Peter	Johnson	1825	Dwight Lagrange	great-grandson
			Earl Lagrange	great-grandson

ANCESTOR	COUNTY	DATE	DESCENDANT	RELATIONSHIP
LAIN, Jacob	Hancock	1843	Hazel (Lain) Leslie	granddaughter
LAKE, William	Fayette	1838	Boyd Lake	great-grandson
LAMBERSON, Thoroughgood	Fayette	1844	Inez Opal Lamberson	great-granddaughter
LAMBERT, James	Hendricks	1837	Florence Brush*	niece
LAMBERT, John	Fayette	1823	Pearl (Rigor) Nothern	great-granddaughter
LAMBERT, John	Hendricks	1837	Charles O. Swaim	great-grandson
LANDES, Frederick	Carroll	1839	Max G. Landes	great-great-grandson
LANE, William	Boone	1835	Mrs. Roy Dye	granddaughter
LANG, William	Johnson	1834	Mrs. W. A. Collett	great-granddaughter
LANKFORD, Robert F.	Marion	1840	Luther E. Esterday	grandson
LATOURETTE, John	Fountain	1828	Fred Cooper LaTourette	grandson
LEACH, Elias	Hendricks	1836	Gertrude Hines	great-granddaughter
LEACH, Enos	Hendricks	1834	John E. Leach	grandson
LEAMING, Furman	Tippecanoe	1844	Curwen Leaming	grandson
LEAMING, Jerry Kellog	Hamilton	1835	Mary (Leaming) Gilkey	granddaughter
LEAVERTON, Anderson	Huntington	1835	Elizabeth C. Kierstead	great-granddaughter
LEE, James	Madison	1835	Helen J. Davis	great-granddaughter
LEGATE, Ivory Holland	Shelby	1829	Susan Keightley	granddaughter
LEGG, Samuel	Rush	1833	L. Meredith Hall	great-grandson
LEITER, Jacob	Fulton	1846	Levi Leiter	son
LEMMON, John S.	Harrison	1846	Walter Lemmon	grandson
LENDERMAN, Henry	Vigo		Laura Lenderman	granddaughter
			Sarah Lenderman	granddaughter
LENNEN, Peter	Hamilton	1830	Iva Lennen	great-granddaughter
LEONARD, George	Grant	1836	Leonard Carroll	grandson
			Merritt Carroll	grandson
			Ruth Carroll	granddaughter
			Waldo Leonard	grandson
LEONARD, Henry	Hamilton	1835	Will Leonard	great-grandson
			Winnie (Leonard) Wibel	great-granddaughter
LESLEY, Peter	Randolph	1819	David J. Lesley	grandson
LESTER, David	Switzerland		William Lester	great-grandson
LEWIS, Harvey	Steuben	1846	H. B. Lewis	son
LEWIS, John	Wayne	1816	Lucile Lewis	great-granddaughter-in-law
LEWIS, Leonard	Fayette	1817	Howard Lewis	great-grandson
LICKING, John F.	Dearborn	1846	Wesley Licking	grandson

Samuel Welch was a soldier of the Revolutionary War. . . . [He] came to Indiana in 1815; built grist mill & dam on land . . . served in Legislature; advocate [for] building a charcoal bed for road thru to Lake Michigan, which is today the Michigan Rd. He is buried on the above farm & has a government marker.

Finley Ralston
Jefferson County

Jacob Sippy is a son of Joseph Sippy who came to America with Gen. Lafayette's fleet to assist in the American Revolution. The soldier was ill with small-pox . . . the fleet sailed home without him. He recovered, assisted Mrs. John Johnson on the coast to care for sick soldiers. . . . Joseph Sippy and mother-in-law were awarded 494 acres of land in Beaver County, Penn., for their service. . . . they came on to Indiana, assisting in building the civilization of our Great Indiana.

Ina Brundige
Fulton County

ANCESTOR	COUNTY	DATE	DESCENDANT	RELATIONSHIP
LIGHT, Abner	Owen	1818	Mrs. Myron Beaman	great-great-granddaughter
			Elizabeth Messick	granddaughter
LIGHT, Hugh M.	Owen	1836	Ione (Light) Figg	great-granddaughter
LINCOLN, Moses Jefferies	Warren	1829	Verlie I. Blue	great-granddaughter
LINDER, George	Dubois	1844	Delphena (Linder) Niehaus	granddaughter
LINDLEY, Aaron	Hamilton	1838	Lyndon K. Beals	great-grandson
LINDLEY, Lot	Parke	1831	Guy Lindley	grandson
LININGER, George	Huntington	1842	Vera B. Stetzel	great-granddaughter
LINK, Thomas Newton	Rush	1839	John Link	grandson
			Mrs. Warner Wyatt	granddaughter
LINN, Andrew	Randolph	1839	Charles A. Linn	great-grandson
			Howard A. Linn	great-great-grandson
LINVILLE, James	Rush	1820	Wilbur Linville	great-grandson
LITCH, Michael	Harrison	1847	William V. Litch	great-grandson
LITTLE, Jehu	Washington	1820	Rose (Wheeler) Lawyer	great-great-granddaughter
LITTLER, Nathan	Grant	1844	Virginia (Richards) Martin	great-granddaughter
			Cleo (Littler) Richards	granddaughter
			Mrs. Mark Richards	granddaughter
LIVEZEY, Isaac	Henry	183?	Lowell H. Smith	great-grandson
LLOYD, Zephaniah	Jefferson		Serilda Lloyd	not given
LOCKYEAR, Christopher	Warrick	1837	Roy Phelps	great-grandson

IHS INDIANA EXTENSION HOMEMAKERS COLLECTION

ANCESTOR	COUNTY	DATE	DESCENDANT	RELATIONSHIP
LOGAN, James	Rush	1821	Mary E. Logan	granddaughter
			Thomas Ora Logan	grandson
LOGAN, Martin	Decatur	1820	Harold L. Beall	great-great-grandson
LOGAN, Thomas Wood	Kosciusko	1838	Ward Thomas Logan	grandson
LOGE, Andreas	Dearborn	1835	Emily K. Hiller	great-granddaughter
LONG, Reuben	Whitley	1836/37	LeNore (Hoard) Enz	great-granddaughter
			Bernice Long	great-granddaughter
LOOMIS, Henry	La Porte	1837	William Loomis	grandson
LOONEY, Peter	Rush	1820	Grace Armstrong	granddaughter
LOOP, Christian	Montgomery	1834	Clyde D. Loop	great-grandson
			Robert Loop	great-great-grandson
			Wilber C. Loop	great-great-grandson
			Mary Louise (Loop) Simms	great-great-granddaughter
LOUGH, David	Fulton	1838	Elizabeth Lough*	granddaughter-in-law
LOWRY, Alexander	Scott	1819	Dennis Cox	great-grandson
LUCAS, Marady	Owen	1837	Theodore Lucas	grandson
LUDWIG, Phillip	Spencer	1850	Jane Ann (Ludwig) Pfender	granddaughter
LUMPKIN, James	Wayne	1832	Elmer Lumpkin	grandson
LUSE, Mathias	White	1837	George C. McCauley	great-grandson
LUTES, John	Jackson	1824	Winfrey Lutes	great-grandson
LUYSTER, Stephen	Johnson	1826	Hugh Hamilton	great-grandson
LYNCH, John	Crawford	1837	Everett Lynch	not given
LYONS, Elijah	Jay	1849	C. Homer Lyons	grandson
MABBITT, John	Carroll	1835	Margaret Mabbitt	great-granddaughter
MACY, James, Sr.	Henry	1834	Henry Wilson Gilbert	grandson
MACY, Tristram B.	Rush	1843	Aldelma Dunning	granddaughter
MADDOX, Frederick	Warren	1829	Ruth Ann White	great-granddaughter
MAGEE, John	Rush	1829	Mrs. Frank Magee*	granddaughter-in-law
MAKEPEACE, Amasa	Madison	1823	Sherman Makepeace	grandson
			Willard Makepeace	great-grandson
MALCOLM, Peter	Vigo	1841	Mrs. Fred M. Goodman*	great-granddaughter-in-law
MALLERY, Horace	Hamilton	1839	Kate Mallery*	granddaughter-in-law
MALSBURY, Jacob	Grant	1841	Mary (Malsbury) Taylor	daughter
MANNAN, John	Owen	1822	Frank Mannan	great-grandson
MARCH, Daniel	Pulaski	1839	Carl Marsh	not given
MARKEY, Benjamin	Montgomery	1834	Mrs. George Markey*	great-granddaughter-in-law
MARKLE, Abraham	Vigo	1816	Ann Markle	great-granddaughter

This land was called canal land and was sold to complete the Canal from the Mouth of the Tippecanoe River to Terre Haute.

Olen York
Marshall County

My great-grandfather, Wm. Bruce was the father of twenty five children. He was the first settler in the village & that is where the village of Bruceville Indiana gets its name. The farm has been in the family for 142 years. His house was the second brick house built west of the Allegheny Mountains & East of the Mississippi river. The first was the Wm. Henry Harrison house in Vincennes.

Cora McClure Goodman
Knox County

ANCESTOR	COUNTY	DATE	DESCENDANT	RELATIONSHIP
MARQUIS, William	Posey	1838	Eva Stevens	daughter
MARSHALL, Aaron	Wayne	1831	Bertha (Marshall) Turner°	great-grandniece-in-law
MARSHALL, John	Fountain	1831	Mary (Marshall) Cox	granddaughter
MARSHALL, Leroy	Vermillion	1835	Carl S. Marshall	grandson
MARSHALL, Samuel	Huntington	1837	Mary A. (Marshall) Pinkerton	daughter
MARSHALL, Thomas	Wayne	1815	Ethel Marshall°	great-great-granddaughter-in-law
MARSHALL, Thomas K.	Jefferson	1825	Lizzie Marshall°	granddaughter-in-law
MARTIN, David	Decatur	1822	Elizabeth (Power) Barnett	great-great-granddaughter
MARTIN, Ephraim	Fountain	1834	Ben C. Martin	not given
MARTIN, James	Carroll	1832	John R. Martin	great-grandson
MARTIN, James	Hendricks	1833	Paul Martin	great-grandson
MARTIN, William	Fayette	1821	Roy Hartman	great-grandson
MARTIN, William M.	Washington	1814	J. C. Bright	great-grandson
			Olive Bright	great-granddaughter
MASON, George	Jackson	1841	Thomas N. Logan	grandson
MASON, James	Sullivan	1823	Harry Mason	great-grandson
MASTEN, John	Hendricks	1833	Walter M. Hodson	great-grandson
			Nellie West	great-granddaughter
MASTEN, Reuben Stanley	Hendricks	1834	Arthur Masten	grandson
MATTHEWS, James	Jefferson	1819	Thomas S. Matthews	great-grandson
MATTIX, Benjamin	Pulaski	1849	Mrs. Wilson Mattix°	granddaughter-in-law
MATTIX, Giles	Clinton	1831	Horace E. Knapp	great-grandson
MAUCK, Jonathan	Harrison	1812	Alva Lang	great-grandson
MAUS, John	Miami	1842	Mrs. Robert Donaldson	great-granddaughter
MAUZY, William	Rush	1840	Chester C. Mauzy	grandson
MAXAM, John S.	Gibson	1835	Carl J. Maxam	grandson
			Loren Maxam	grandson
MAY, David	Monroe		Clara B. May	great-granddaughter
MAYLEY, John	Shelby	1847	Walter Wertz	grandson
MCALISTER, Zachariah	Washington	1811	Jesse Mahuron	great-great-grandson
MCCAIN, Francis	Carroll	1829/30	Burton A. McCain	grandson
			Jess McCain	grandson
MCCARTY, Aaron Hardridge	Owen	1827	George McCarty	great-great-grandson
			Jerry McCarty	great-grandson
MCCARTY, William	Rush	1828	Paul S. Rich	not given
MCCLAMROCK, James	Montgomery	1847	Robert McClamrock	grandson
MCCLELLAN, Starrett	Knox	1844	Estelle Emison	granddaughter
MCCLINICK, ——	Clark	1812	Alpha F. Gilmore	great-great-grandson

ANCESTOR	COUNTY	DATE	DESCENDANT	RELATIONSHIP
MCCLINTIC, Esten	Kosciusko	1834	Charles F. McClintic	grandson
			Martin V. McClintic	grandson
MCCLURE, Daniel	Knox	1804	Maude (Brentlinger) Hohn	great-great-great-granddaughter
MCCONNEL, Sampson	Decatur	1823	Aimee R. McConnell*	granddaughter-in-law
MCCORD, David	Knox	1831	William Purcell Lester	great-great-grandson
MCCORD, James	Warren		Hannah E. Anderson	great-granddaughter
MCCORMICK, Jane	Montgomery	1832	Wilbur Spencer	grandson
MCCOY, Andra	Decatur	1823	John A. McCoy	grandson
MCCOY, Angus C.	Decatur	1823	Eugene M. McCoy	great-grandson
MCCOY, George S.	Decatur	1830	Charles McCoy	grandson
MCCOY, William	Rush	1846	McCoy Carr	grandson
MCCRACKEN, Richard	Daviess	1823	R. W. McCracken	grandson
MCCREARY, Thomas	Switzerland	1840	Addie McCreary	grandson
MCCRORY, John	Fayette	1820	Janet Alexander	great-granddaughter
			Margaret Alexander	great-granddaughter
MCCRORY, Samuel	Rush	1829	John M. Cregor	great-great-grandson
MCCULLOUGH, Samuel	Wayne	1828	Carroll B. McCullough	not given
			Harry Moore	great-great-grandson
MCDILL, Samuel	Union	1815	Mark McDill	great-grandson
MCDOWELL, James	Carroll	1826	John L. McDowell	grandson
MCFARLAND, Demis	Marion	1823	Bertha G. Duzan	granddaughter
MCGOVERN, Philip	Martin	1839	Cecelia Arvin	great-granddaughter
MCGUIRE, Addison	Whitley	1838	Fannie (McGuire) Fry	granddaughter
MCKINNEY, Green	Pike		James McKinney	grandson
MCKINNEY, Prestly T.	Fountain	1837	Glenn E. McKinney	grandson
MCKINNEY, William V.	Clinton	1835	Mrs. Bert McKinney*	granddaughter-in-law
MCKINNIS, Philip	Warren	1843	Mrs. William McKinnis*	daughter-in-law
MCKNIGHT, Elijah	Lawrence	1846	Elpha (McKnight) Fishback	daughter
MCKOWN, Archibald	Henry	1832	Etta Frazier	great-granddaughter
MCLANE, John	La Porte	1832	George L. McLane	grandson
MCLAUGHLIN, George	Decatur	1845	Frances (McLaughlin) Shirk	granddaughter
MCMAHON, Peter	Whitley		Katie (Geiger) Ort	granddaughter
MCMANAMAN, James	Decatur	1845	Claudia (Hill) Alexander	great-granddaughter
MCRAE, Alexander	Harrison	1837	Robert E. Kirkham	great-grandson
MCWHORTER, Tyler	Franklin	1814	Loren McWhorter	great-grandson
MEAL, George	Rush	1825	Chester Meal	great-grandson
MEDLIN, John	Boone	1832	Edith O. Titus	granddaughter
MEEK, Adam	Decatur	1816	Homer G. Meek	great-grandson

This family [Cornthwaite] own also a detached parcel of land located on the Wabash River, the north boundary of which marks the beginning of the Ten Oclock Line established by treaty with the Indians in pioneer days.

Homer G. Cornthwaite
Parke County

As I know the story Henry Ashby received this farm as a gift from my great-grandfather Ashby when he married my grandmother. . . . My great-grandfather was one of the first settlers in Pike County.

Paul W. Ashby
Pike County

IHS INDIANA EXTENSION HOMEMAKERS COLLECTION

ANCESTOR	COUNTY	DATE	DESCENDANT	RELATIONSHIP
MEHARRY, James	Montgomery	1827	Annie V. Meharry	granddaughter
			Ira G. Meharry	grandson
			Roy H. Meharry	great-great-grandson
MEHARRY, Thomas	Tippecanoe	1835	Lyman N. Maddux	great-great-grandson
MELVIN, James M.	Warren	1836	William Melvin Clawson	grandson
MEREDITH, Robert	Fulton	1836/38	Mrs. Roy Adamson	granddaughter
			Silas M. Meredith	grandson
MERRIMAN, William	Morgan	1835	Orville R. Wells	great-grandson
MESSMORE, Andrew	Fountain	1850	Pearl Moser	granddaughter
METZGER, John	St. Joseph	1832	John T. Metzger	grandson
MEYERS, John, Sr.	Rush	1820	Mabel Norris	great-granddaughter
			Mildred Wells	great-granddaughter
MEYERS, Nicholas	Miami		Edna Bowman	great-granddaughter
MILAM, Joel	Wabash	1844	Peter J. Milam	grandson
MILER, Thomas	Crawford	1845	Mrs. Edward Faulkenburg	great-granddaughter
MILLER, Aaron	Rush	1831	Grant Miller	great-grandson
MILLER, Daniel	Ripley	1832	Charles L. Roepke*	grandnephew
MILLER, George	Decatur	1824	Charles Ira Miller	great-grandson
MILLER, John	Hendricks	1825	Burke H. Miller	grandson
MILLER, John	Clay	1838	Elsie Mae Phillips*	great-grandniece
MILLER, John	Wabash	1841	Nellie D. Reahard	great-granddaughter
MILLER, John B.	Rush	1834	Lorrie Miller	great-grandson

Hay barn in Miami County

IHS INDIANA EXTENSION HOMEMAKERS COLLECTION

ANCESTOR	COUNTY	DATE	DESCENDANT	RELATIONSHIP
MILLER, John D.	Hendricks	1833	Nellie (Miller) Little	granddaughter
MILLER, John J.	Spencer		Flora Gabbert	great-granddaughter
MILLER, John Jacob	Harrison	1814	Thomas O. Miller	great-grandson
MILLER, John W.	Wayne	1806	John W. Miller Jr.	great-great-grandson
MILLER, Peter	Harrison	1839	Elizabeth Quebbeman	granddaughter
MILLER, Robert	Porter	1850	Eugene S. Miller	son
			Joe Robbins	grandson
			Keneth Wolf	great-grandson
MILLIGAN, Wilson	Jay	1837	Homer Milligan	great-grandson
			Milo Milligan	great-grandson
MILLS, Anderson B.	Owen	1836	Denver Gibson	great-grandson
MILN, William	Gibson	1833	Conrad M. Howe	great-grandson
MINICK, Justus	Cass	1844	Alvin E. Minnick	great-grandson
MINTS, William	Hancock	1836	Salome D. Mints	granddaughter
			William T. Mints	grandson
MITCHELL, John D.	Johnson	1849	Mrs. Emmett Pritchard*	great-granddaughter-in-law
MOCK, Aaron	Rush	1840	Aaron Kennedy	grandson
MODESITT, Nathaniel Hardester	Clay	1835	Annie M. Modesitt	granddaughter
			Ruth L. Modesitt	granddaughter
MOFFATT, John, Sr.	Bartholomew	1835	Harvey Moffat Keller	great-grandson
MOFFITT, John	Jefferson	1835	Harvey Moffitt Keller	not given
MONEY, Robert	Shelby	1822	Nannie Money*	granddaughter-in-law

Jesse Nash knew George Rapp in Penn. and was influenced by him to move his family to New Harmony. His family and belongings came on a flatboat to Mt. [Vernon?] and there he was met by George Rapp who helped him move on to the settlement of Harmony.

Eugene W. Nash
Posey County

The old homestead still remains on the original forty acres. To this has been added in later years, adjoining land, the farm now contains 228 acres.

Mrs. Wilson Mattix
Pulaski County

ANCESTOR	COUNTY	DATE	DESCENDANT	RELATIONSHIP
MONEY, William	Jay	1837	W. F. Hilfiker	grandson
MONTGOMERY, James	Daviess	1823	Mrs. Cletus J. Montgomery*	great-granddaughter-in-law
MONTGOMERY, James	Cass	1837	James R. Montgomery	not given
MONTGOMERY, Richard	Jackson	1820	T. Harlan Montgomery	great-grandson
MOON, Nelson	St. Joseph	1842	Chester Newman	great-grandson
MOORE, Artemas	Boone	1837	Anson M. Bell	grandson
MOORE, Dennis	Fulton	1836	Robert R. Burns	great-grandson
MOORE, James	Clinton	1838	Sarah Grace Smith	granddaughter
MOORE, James	Rush	1842	Henry Chase Moore	grandson
MOREHOUSE, Wilbur	Jay	1839	Mrs. W. C. Bailey	granddaughter
MORGAN, Jacob	Warren	1837	Mrs. W. O. Smith	great-granddaughter
MORGAN, Jesse	Porter	1835	Bennett Morgan	grandson
			Edward L. Morgan	grandson
MORGAN, John	Franklin	1846	Mrs. John LaMont	great-granddaughter
MORGAN, Randall C.	Knox	1824	Leslie Morgan	grandson
MORRIS, Burr	Martin	1844	Frank H. Morris	grandson
MORRIS, Jehoshaphat	Washington	1825	Emory V. Morris	great-grandson
MORRISON, Daniel	Cass	1842	Mrs. Dudley Bridge	great-great-granddaughter
MOSER, Michael	Putnam		W. M. Moser	grandson
MOSHER, Benona	Whitley	1839	Herma (Mosher) Price	great-granddaughter
MOSS, John	Madison	1837	R. Wysong Julius	great-great-grandson

Decatur County

IHS INDIANA EXTENSION HOMEMAKERS COLLECTION

ANCESTOR	COUNTY	DATE	DESCENDANT	RELATIONSHIP
MOSS, William	Union	1808	Gertrude L. McKee	great-granddaughter
			Cora (Hart) Stevens	great-granddaughter
MOSSLER, John G.	Harrison	1846	J. E. Mossler	grandson
MOUNSEY, John	Wells	1840	George Rollin Osborn	grandson
MOYER, George	Tippecanoe	1826	Ralph C. Foutz	great-great-grandson
MULLENDORE, Jacob	Shelby	1832	India (Mullendore) Hartman	granddaughter
MULLIN, Noah	Carroll	1834	Herman C. Mullin	grandson
			Roscoe C. Mullin	great-grandson
MULLIS, Jacob	Lawrence	1836	Asher B. Voyles	great-grandson
MUMFORD, Thomas	Posey	1847	Thomas F. Mumford	great-grandson
MUNCY, John	Boone	1834	Brewer Demaree	great-grandson
MUNDY, James	Hamilton	1847	Kenneth Clark Mundy	great-grandson
MUNGER, Edwin	Fayette	1821	Warren H. Munger	grandson
MURPHY, James	Posey	1829	Robert Alex Lamar*	great-grandson-in-law
MURPHY, James	Rush	1835	Mrs. Frank J. Murphy*	granddaughter-in-law
MURPHY, John	Fayette	1814	Harlie Foster	great-granddaughter
MURPHY, John	Miami	1846	Earl Murphy	grandson
MYER, John, Sr.	Carroll	1834	Royce Myer	great-grandson
NASH, Andrew	Posey	1840	Eugene W. Nash	son
NASH, Jesse	Posey	1829	Eugene W. Nash	grandson
NEIDEH, Solomon	DeKalb	1838	Edson Klinkel	great-grandson

The house built by John David Dufour still stands. It is a typical Swiss house with long galleries (or porches) across the front and back. My home was built by my father, T. R. Dufour, during the Civil War.

Olive D. Trafelet
Switzerland County

> **The original homestead, where Joseph Butler was born, still stands, though in a bad condition. We have been told that it is the oldest house in northern Indiana.**
>
> Myrtle R. Butler
> *Steuben County*

ANCESTOR	COUNTY	DATE	DESCENDANT	RELATIONSHIP
NELSON, Anthony	LaGrange	1832	Edith Mast	great-granddaughter
			Keith Smith	great-grandson
			Wilma Smith	great-granddaughter
			Ramah Strombeck	great-granddaughter
NEUNAN, Nichles B.	LaGrange	1836	H. B. Lewis	grandson
NEWBY, Henry	Rush	1833	Sarah Newby	granddaughter
NEWBY, Micajah	Grant	1837	Hanley Thomas	grandson
NEWHOUSE, Levi J.	Rush	1847	Oscar Newhouse	grandson
NEWKIRK, Richard	Jackson	1835	John E. Bergdoll	great-grandson
NEWLAND, James	Fayette	1818	Elmer Newland	great-grandson
NEWLIN, Jonathan	Orange	1819	Mort Newlin*	grandnephew
NEWMAN, Gary P.	LaGrange	1838	Mrs. William Walter	granddaughter
NEWMAN, Issac	Miami		Zola Andrews	granddaughter
			Mrs. Omer Maus	granddaughter
			Charles Newman	grandson
NEWMAN, John	Hendricks	1834	Stella Shields	great-granddaughter
NEWSOM, Willis	Bartholomew	1820	Hadley C. Thomas	great-grandson
NICEWANDER, Joseph	Tippecanoe	1831	Harry Nicewander	grandson
NIEDERHAUS, Jacob	Gibson	1837	William Niederhaus	grandson
NIEHAUS, Henry, Sr.	Dubois	1846	Henry Niehaus	grandson
			John B. Niehaus	grandson
NIEWALD, William	Knox	1847	Paul Niewald	grandson
NILES, John Barron	La Porte	1834	Fanny (Scott) Rumely	granddaughter
NIXSON, John	Jay	1836	Joseph R. Nixson	grandson
NOBLE, John Pickett	Porter	1837	C. C. Gregg*	not given
NORRIS, Benjamin	Rush	after 1834	Rema May Fridlin	great-granddaughter
			Helen Martin	great-granddaughter
			Lowell Norris	great-grandson
			Walter Norris	great-grandson
NORRIS, Ransom H.	Marshall	1846	Everett Norris	grandson
NOTESTINE, Peter Nicholas	LaGrange	1846	Herman Notestine	grandson
NUSSEL, Anna Barbera Fleshman	Clay	1839	Adam Nussel	grandson
O'BRIEN, Peter	Martin	1823	Desmond O'Brien	great-great-grandson
ODLE, William	Randolph	1837	Elmer Odle	great-grandson
ODOM, Henry	Henry	1839	Lamont O'Harra*	great-grandnephew
OFFICER, James	Jefferson	1816	Betty Officer	granddaughter
OFFUTT, Sabert S.	Rush	1839	Denning Nelson	great-grandson

ANCESTOR	COUNTY	DATE	DESCENDANT	RELATIONSHIP
O'HARA, Michael	La Porte	1828	Arden L. Hunt°	grandnephew
O'HARA, William	La Porte	1837	George G. O'Hara	son
OHLEMEYER, George Wm.	Clay	1837	Lafe B. Ohlemeyer	great-grandson
OLDHAM, James	Rush	1847	Lydia (Oldham) Sample	great-granddaughter
OLDHAM, Samuel	Rush	1847	Lydia (Oldham) Sample	granddaughter
OLMSTEAD, Samuel Laird	Vanderburgh	1836/37	Mrs. H. Beach	granddaughter
			Marshall Olmstead	not given
OLNEY, John	LaGrange	1830	Blanche (Olney) Myers	great-granddaughter
OREN, Jesse	Grant	1841	Warren Oren	grandson
OSBORN, Aaron	Switzerland	1812	C. L. Jackson	grandson
OSBORN, Charles	Wayne	1819	Daisy Osborn	great-granddaughter
OSBORN, Jonathan	La Porte	1834	Dotha L. Osborn°	granddaughter-in-law
OSBORN, Jonathan, Jr.	La Porte	1863	Bonnie (Osborn) Hixon	daughter
OSBURN, William	Sullivan	1835	Josie Conner	granddaughter
			Chancey Osburn	grandson
			H. T. Osburn	grandson
			Jessie Osburn	granddaughter
			Mary F. Osburn	granddaughter
OVERLEESE, Daniel	Rush	1820	Warder B. Julian	great-great-grandson
OVERMAN, William	Miami	1845	Louie (Overman) Douglass	granddaughter
			Herman Overman	grandson

This farm has always remained in [the Gray] family. My son . . . and his two sons . . . also live on this farm, which makes the sixth generation. The information has been handed down from generation to generation that the Lewis Clark Expedition crossed Big Sugar Creek on this farm.

Mary Smith Gray
Shelby County

HISTORY OF RUSH COUNTY, 1888

Benjamin F. Norris, original owner of Centennial Farm in Rush County. See entry for Norris on opposite page.

There has never been a mortgage on this land & taxes have been paid promptly. The redbrick house in which my Mother Sallie Legate Keightley was born July 16 1843 . . . is in splendid state of preservation.

Susan Keightley
Shelby County

ANCESTOR	COUNTY	DATE	DESCENDANT	RELATIONSHIP
OWINGS, George	Grant	1837	Louis L. Needler	grandson
			Oma Schweitzer	granddaughter
OZBURN, Isaac	Randolph	1845	Jonathan M. Ozburn	son
PACE, Uriah	Delaware	1842	Ralph W. Pace	grandson
PALMER, ——	Wayne	1841	L. C. Palmer	not given
PARK, David	Johnson	1835	Omer Park	grandson
PARK, Esther	Johnson	1834	Mrs. C. P. Vandivier	granddaughter
PARKER, John	Clinton	1833	Mary C. Lane*	third cousin
PARKER, John L.	Hendricks	1837	Lloyd W. Hendrickson*	great-great-grandnephew
PARSLEY, James	Spencer		Mrs. Howard Parsley*	great-granddaughter-in-law
PARTRIDGE, Edwin	Fulton	1843	Francis Partridge	grandson
			Mina Partridge	granddaughter
			Olive Partridge	granddaughter
PATTERSON, Amassa	Rush	1841	Howard Winslow	great-grandson
PATTERSON, Thomas	Johnson	1834	Chester Beckley	grandson
PATTISON, Edward	Rush	1823	Elizabeth (Pattison) Schrichte	great-granddaughter
PATTY, James, Sr.	Carroll	1835	Walter E. Patty	grandson
PAUL, George	Vanderburgh	1841	Joseph C. Brune	great-grandson
			Rose Brune	great-granddaughter
			Severina J. Martin	great-granddaughter
PAULUS, Henry	Elkhart		Ruth (Paulus) Eby	granddaughter

IHS INDIANA EXTENSION HOMEMAKERS COLLECTION

[Levi J. Newhouse] planted his first crop of corn with a spade. In between where the sun would shine in. Drove his first bunch of hog[s] to Cincinnati, his closest market.

Oscar Newhouse
Rush County

ANCESTOR	COUNTY	DATE	DESCENDANT	RELATIONSHIP
PAYTON, Arthur	Vermillion	1832	Julia (Jones) James	great-great-granddaughter
PEACHER, William	Orange	1843	Cecil Peacher	great-grandson
			Harry Peacher	grandson
PEACOCK, Amos	Randolph	1838	Ralph C. Peacock	great-great-grandson
PEARCE, Milton	Warren	1837	Mrs. W. O. Smith	granddaughter
PEARSON, William	Parke	1828	Allen H. Pearson	great-grandson
PECK, Samuel	Clark	1825	John L. Gibson*	great-grandnephew
PEEK, Cager	Martin	1836	R. M. Peek	grandson
PENCE, Jacob	Grant	1845	Benjamin F. Pence	son
PENCE, Peter	Parke	1835	Floyd Clark	grandson
			Walter Clark	grandson
PETER, William	Clinton	1833	Laura (Peter) Landes	great-granddaughter
PETER, William	Tippecanoe	1835	Lois E. (Peter) Henderson	great-granddaughter
PETERS, Michael	Switzerland	1828	Harry C. Peters	grandson
PETTYJOHN, William	Clark	1815	William Dickson	great-grandson
PFENDLER, Nickolas	Shelby	1847	David C. Perdue	great-grandson
			Robert Pfendler Perdue	great-grandson
			Thomas S. Perdue	great-grandson
PFENNING, Conrad	Jackson	1848	Edgar H. Stahl	great-grandson
PFRIMMER, Samuel	Harrison	1815	Clara (Pfrimmer) Hays	great-granddaughter
PICKEL, Joseph	Knox	1841	Grace Shaw*	grandniece
PIEPER, Anton	Knox	1847	Mrs. Ernest W. Tilly	granddaughter

My wife Laura Klingensmith Artman has lived here since May 1868. We have grease lamp, candle molds and use of the Old Dutch oven for baking, and the fire place and never moved only from old house into the new in 1896. Now use electricity, gas for heating and cooking.

Wm. A. Artman
Marion County

ANCESTOR	COUNTY	DATE	DESCENDANT	RELATIONSHIP
PIPER, Josiah	Fayette	1824	Mrs. Minor Thomas	great-great-granddaughter
PIRTLE, Abel	Sullivan	1835	W. C. Monroe	grandson
PITMAN, David	Harrison	1816	John R. Pitman	great-grandson
PITMAN, David	Harrison	1829	Joe A. Pitman	great-great-grandson
PITTENGER, John	Delaware	1836	Lola Thornburg	great-granddaughter
PLATT, Abraham	Randolph	1837	Owen Platt	grandson
PLEAK, Fielding	Decatur	1831	Elizabeth (Pleak) Kanouse	great-granddaughter
			Willamette (Pleak) Lemmon	great-granddaughter
PLUMMER, Jackson	Shelby	1846	Merrill Plummer	grandson
PLUMMER, Joseph	Hendricks	1832	O. E. Plummer	grandson
POLK, Charles	Perry	1807	Judd Polk	not given
PORTER, Alexander	Decatur	1837	Andrew Porter	grandson
PORTER, Benjamin	Cass	1827	Albert Porter	grandson
PORTER, John	Boone	1831	Marvin M. Porter	not given
PORTER, Joshua	Fayette	1811	Nelson Porter	great-grandson
PORTER, Natham	Rush	1831	Anna Sipe	granddaughter
POTE, Thomas	Posey	1846	Jennie (Kemmerling) Schmidt	granddaughter
POTTER, Silas R.	Switzerland	1845	Edward H. Potter	grandson
POTTROFF, Andrew	Jackson	1853	Evangeline (Stewart) Booker	great-granddaughter
POTTS, Henry	Harrison	1830	Anna (Lamb) Hyndman*	great-grandniece
			Wilbur G. Lamb*	great-grandnephew

HISTORY OF RUSH COUNTY, 1888

John Arnold, M.D., original owner of Centennial Farm in Rush County. **See entry for Arnold on page 43.**

ANCESTOR	COUNTY	DATE	DESCENDANT	RELATIONSHIP
POTTS, Joseph	Harrison	1800/10	Anna (Lamb) Hyndman	great-granddaughter
			Wilbur G. Lamb	great-grandson
POUND, Hezekiah	Scott	1817	Imogene (Pound) Smith	great-great-granddaughter
POWELL, Harrison	Grant	1839	Blanch (Mart) Swaim	granddaughter
POWELL, Isaac	Fayette	1833	Homer Powell	grandson
PROEBSTER, John Adam	Hendricks	1832	Cecil F. Prebster	great-grandson
PUCKETT, Thomas	Vigo	1816	Ernest L. Smith	great-grandson
PUGH, Jefferson	Grant	1848	Emma Anthony	granddaughter
PUGH, Michael	Grant	1836	Paul D. Pugh	great-grandson
PULLEY, Samuel, Sr.	Grant	1839	Nellie Lucille (Pulley) Blue°	great-granddaughter-in-law
			Margaret Alice Pulley°	great-great-granddaughter-in-law
QUINN, John	Union	1827	Ruby M. Parks	great-granddaughter
			Charles F. Winters	great-grandson
QUINN, William, Sr.	Carroll	1837	Lottie (Quinn) Seward	great-granddaughter
RADCLIFFE, Thomas	Fountain	1828	Capitola (Ratcliffe) Lindley	great-granddaughter
			Cedella Ratcliffe	great-granddaughter
RADER, Adam	Wayne	1834	Esther (Boroughs) Riggs	great-granddaughter
RAGSDALE, Frederick	Johnson	1836/39	Ruth M. Jacobs	great-granddaughter
			Clarence D. Ragsdale	great-grandson

Two great grandchildren of James H. Stewart are participating in the first "Pioneer Glory" historical pageant of Indiana. . . . My grandfather [James H. Stewart] . . . owned 1000 acres at the time of death.

Evangeline Stewart Booker
Jackson County

IHS INDIANA EXTENSION HOMEMAKERS COLLECTION

Sarah Jane Clifford in her Fayette County cabbage patch, 1908

This farm is still in my possession. The original house is still standing in good repair. Built in 1824. We have the papers to prove these statements.

Fred H. Brookbank
Union County

IHS BASS PHOTO CO. COLLECTION, #31:4

ANCESTOR	COUNTY	DATE	DESCENDANT	RELATIONSHIP
RALSTON, Matthew	Jefferson	1830	Walter Ralston	grandson
RALSTON, William	Jefferson	1825	Mabel Ralston*	great-great-grandniece
RAMER, Henry	Cass	1842	Charles L. Ramer	great-grandson
RAMSEYER, Jacob	Switzerland		Omer Ramseyer	great-grandson
RANCK, George	Wayne	1830	Ralph Ranck	great-grandson
RAY, John	Vigo	1833	Laura Ray*	granddaughter-in-law
RAY, John	Marshall	1835	George C. Ray	grandson
REA, David	Ripley	1838	Robert S. Rea	grandson
REA, Robert R.	Jefferson	1833	John Hutchison	not given
			Margaret Hutchison	not given
REAVIS, William, Sr.	Gibson	1837	William C. Reavis	great-grandson
REDMAN, William	Gibson	1815	Mrs. Prentice O. Williams	great-granddaughter
REED, B. F.	Rush	1830	Harry Armstrong	grandson
REED, William	Harrison	1818	Aldine Snyder	great-great-granddaughter
REES, David	Dearborn	1811	Loren Stratton Rees	great-grandson
			R. Hamilton Rees	great-grandson
REID, Lewis	Bartholomew	1833	Ralph Heilman	great-grandson
REITENOUR, George	Randolph	1819	Pauline (Reitenour) Harshman	not given
			William A. Reitenour	great-grandson
			Flora C. Wine	great-granddaughter
REUTMANN, Charles J.	Dubois	1840	John Reutmann	grandson
			William Reutmann	grandson

ANCESTOR	COUNTY	DATE	DESCENDANT	RELATIONSHIP
REX, John	Clinton	1839	Clarence C. Rex	grandson
REYNOLDS, John	White	1834	Lottie Reynolds	granddaughter
REYNOLDS, Samuel	Wayne	1843	Wallace Reynolds	great-grandson
RHOAD, Philip	Hamilton	1837	Cecil Rhodes	grandson
			Rex Rhodes	grandson
RHODE, Jonathan	Warren	1828	James N. Rhode	grandson
RHODE, William	Warren	1830	Cora J. Rhode	great-granddaughter
			J. Clay Rhode	great-grandson
			Lillis Rhode	great-granddaughter
			Elsie J. Smith	great-granddaughter
RHODES, Daniel	DeKalb	1834	Mirwood E. Rhodes	great-grandson
RHYAN, Henry	Vigo	1830	Sherman L. Rhyan	grandson
RICE, Ebenezer	Scott	1827	Mrs. Charles P. Harrison	granddaughter
RICH, Peter	Hamilton	1838	W. J. Horney	great-grandson
RICHARD, Henry	Harrison	1843	Frank W. Shireman	grandson
RICHERT, Jacob, Sr.	Harrison	1838	Jess A. Richert	great-great-grandson
RICHEY, John	Boone	1834	Walter Richey	grandson
			Mildred Tomlin	great-granddaughter
RICHWINE, Joel	Madison	1837	Carrie Allen	granddaughter
			Joel M. Cox	grandson
RIDDLEBARGER, David	Randolph	1837	George Lewis	great-grandson
			Ethel (Reitenour) Muchow	great-granddaughter
RIDGEWAY, Jeremiah	La Porte	1837	John A. Ridgeway	great-grandson
RIDGWAY, James	La Porte	1831/47	Pearl R. Grandstaff	granddaughter
			John A. Ridgway	grandson
RIGGS, Ransom, Sr.	Johnson	1835	William V. Riggs Sr.	son
RINGER, Joseph	Marion	1849	David R. Johnson	grandson
RINGGENBERG, Christ	Marshall	1845	Samuel L. Ringgenberg	son
RINKER, John	Delaware	1842	Atlee Rinker	great-grandson
			Edward L. Rinker	great-grandson
RISINGER, Samuel	Ripley	1837	Laura E. Knauber	granddaughter
RITCHEY, John	Tippecanoe	1832	Rainey McCoy	grandson
			Nancy (Ritchey) Moore	granddaughter
ROBB, David	Gibson	1839	Clarence F. Hitch	grandson
			Othniel Hitch	grandson
ROBB, James B.	Gibson	1802	Clarence F. Hitch	great-grandson
			Othniel Hitch	great-grandson
ROBBINS, Samuel P.	Porter	1835	Lewis H. Robbins	great-grandson
ROBERTSON, Blaze	Jackson	1835	Royal Robertson	grandson

Enoch Russell came from Fayette Co. He hunted & trapped in Wabash Co. as early as 1812. He died in 1849 and was first person buried in Hopewell Cemetery, that is a part of the Gene Stratton-Porter farm in Wabash Co.

Hugh V. Cleaveland
Wabash County

[The] wife of the second owner, Jacob Harvey, was first white female born in Wells Co; Jacob's brother, John; was first white male born there.

Henry H. Harvey
Wells County

IHS BASS PHOTO CO. COLLECTION, #238666

ANCESTOR	COUNTY	DATE	DESCENDANT	RELATIONSHIP
ROBERTSON, Daniel D.	St. Joseph	1842	Myrtle (Ullery) Gillin	granddaughter
ROBERTSON, John	Jackson	1820	Will R. Isaacs	great-grandson
ROBERTSON, Lewis	Carroll	1835	Lula Robertson*	granddaughter-in-law
ROBERTSON, Middleton	Jefferson	1811	W. E. Robertson	grandson
ROBINSON, Elijah G.	Daviess	1840	Arthur B. Clark	grandson
ROBINSON, John M.	Decatur	1820	Martha T. Pleak	not given
ROBINSON, Lewis	Fayette	1825	D. S. Robinson	grandson
ROBINSON, Wiley	Monroe	1847	Maud E. Robinson	granddaughter
ROBY, Clark	Union	1837	S. C. Greene	grandson
ROCKHILL, Hampton	Wabash	1837	Harvey F. Hanley	grandson
RODARMEL, Abraham	Knox	1814	Harriett R. Williams	granddaughter
RODIBAUGH, Samuel	Marion	1830	Russell Hightshue*	grandson-in-law
ROE, Abraham	Randolph	1847	Doris Wright	great-granddaughter
ROEGER, John	Jackson	1845	Charles A. Roeger	son
ROGERS, Burgeas	Putnam	1834	Bertha Rogers	great-granddaughter
			Eston Rogers	great-grandson
			Lemal Rogers	great-grandson
ROGERS, Johnathan	Vigo	1837	Edgar Smith	great-grandson
ROGERS, Joseph	Madison	1834	Clarence Rogers	great-grandson
ROHRBAUGH, Solomon	Delaware		Orlan L. Strong*	grandnephew
ROLL, Isaac	Carroll	1833	Alice (Kennard) Cheadle	great-granddaughter
RONALD, ——	Fayette	1820	Minnie J. Gray	granddaughter

Shocks of grain

IHS BASS PHOTO CO. COLLECTION, #9437

ANCESTOR	COUNTY	DATE	DESCENDANT	RELATIONSHIP
ROOT, Simeon	Knox	1825	Allen A. Root	grandson
ROOT, William	Lawrence	1836	Maude Henderson	granddaughter
ROSE, Martin	Knox	1808	George M. Patterson	great-grandson
ROSS, George	Kosciusko	1848	Charles W. Ross	grandson
ROSS, Jesse	Harrison		Jesse O. Fox	grandson
ROSS, Thomas	Rush	1829	Lora Waggoner	granddaughter
ROSSELL, Jesse	Sullivan	1816	Floyd Hunt	great-grandson
			Raymond Hunt	great-grandson
			Ethel Jones	great-granddaughter
ROTH, Anna (Souder) Witmer	Allen	1844	John Roth	grandson
ROTHERT, Christian	Dubois	1841	Harvey Rothert	grandson
ROUTH, James	Montgomery	1844	Earl Otterman°	grandnephew
RUDDELL, Ambrose G.	Marion	1837	Frank S. Ruddell	grandson
			James H. Ruddell	great-grandson
			Warren T. Ruddell	great-grandson
RUIMER, Daniel	Daviess	1825	George R. Eagle	great-grandson
RULLMAN, Harman Henry	Dearborn	1846	Mrs. William Knollman	granddaughter
RUNNER, Isaac	Benton	1843	Christel (Runner) Halgren	granddaughter
RUPEL, David	St. Joseph		Prudence Jackson	not given
RUSH, B.	Lawrence		John Rush Holmes	grandson
RUSK, David	Fountain	1832	W. E. Rusk	grandson
RUSK, David, Sr.	Montgomery	1830	Edith (Rusk) Runyan	great-granddaughter

This farm has been owned & occupied by members of the [Brouse] family, since [1846]. . . . At present my daughter, (Mrs. Emerson Bushong, her husband & family) reside on the farm and celebrated Sept. 10th, being able to turn on electricity (R.E.M.C.). This makes 5 generations living on same farm.

Mabel Hontz
Noble County

Jonathan Osborn, Sr.—brought his family from southern Indiana in a lumber wagon to their new home. . . . The house was built back from the main road 30 rods—I have been told that many early settlers did this to be nearer all the fields to be cultivated. The house had one large room—at one end was the fireplace with a large crane to hold kettles for cooking—The [parents'] bed stood in the opposite corner—and a narrow winding stairway led to the Loft—where the seven boys slept.

Bonnie Osborn Hixon
La Porte County

ANCESTOR	COUNTY	DATE	DESCENDANT	RELATIONSHIP
RUSSELL, Enoch	Wabash	1835	Hugh V. Cleaveland	great-grandson
RUSSELL, Jacob, Sr.	St. Joseph	1838	Leo R. Metzger	great-grandson
RUSSELL, James	Jackson	1822	Bertha Bryan	great-granddaughter
RUSSELL, John	Vermillion	1836	Hortense M. Clem	great-great-granddaughter
RYAN, Michael Maher	Daviess	1840	Joseph A. Ryan	grandson
RYBURN, John	Rush	1821	Grant Hinchman	grandson
RYKER, John G.	Jefferson	1805	Edgar Ryker	grandson
RYNER, Henry	Hendricks	1832	Francis H. Ryner	grandson
SANDERS, William	Grant	1834	Fred Sanders	grandson
SANFORD, Edward	Union	1836	Willis M. Sanford	grandson
SARBER, George	Kosciusko	1842	Elmer D. Sarber	great-grandson
SARTOR, James	Knox	1841	Edith (Sartor) Phipps	granddaughter
SATER, Henry	Franklin	1813	Alta B. (Sater) Gurr	great-granddaughter
SATTERTHWAITE, Joel	Huntington	1837	Charles Corwin Satterthwaite	great-grandson
SATTISON, Johnathan	Whitley	1840	Charles Sattison	grandson
SAWDON, William	Dearborn	1835	Otto Sawdon	grandson
SAYERS, Robert Floyd	Tippecanoe	1835	Allie S. Amos	granddaughter
SCHAFER, Conrad, Sr.	St. Joseph	1847	Clem L. Schafer	great-grandson
SCHERMERHORN, Ernestus	LaGrange	1839	John H. Schermerhorn	grandson
SCHLACHTER, Joseph	Dubois	1842	Leo Schlachter	not given
SCHLACHTER, Joseph, Sr.	Vanderburgh	1842	Leo J. Schlachter	grandson
SCHLOSSER, Elias	Warren	1838	Harold Schlosser	great-grandson
SCHMIDLAPP, John David	Jefferson	1838	Edward Schmidlapp	grandson
SCHOLL, Jacob, Jr.	Fayette	1833	Elmer Scholl	grandson
			J. Edgar Scholl	great-grandson
SCHOONOVER, James	Warren	1839	James C. Schoonover	great-grandson
SCHRAMM, Jacob	Hancock	1835	Otto Schramm	grandson
SCHRANTZ, John	Wabash	1843	W. F. Clupper	grandson
SCHULER, Robert	Wabash	1839	Mrs. Donald Dolby	great-granddaughter
SCOTT, John	Wayne	1812	May M. Howe*	great-grandniece
			Jeannette (Howe) Wilson	granddaughter
SCOTT, John	Jefferson	1822	Donald Scott	great-grandson
SCOTT, Tandy	Hendricks	1846	Benjamin Scott	grandson
SEIBERT, Martin	Jackson	1844	Louise (Seibert) Stanfield	granddaughter
SELBY, Otho	Grant	1837	Victor A. Selby	grandson
SELLARS, Henry	Martin	1847	Homer Sellars	grandson

ANCESTOR	COUNTY	DATE	DESCENDANT	RELATIONSHIP
SELLER, J. W. P.	Parke	1845	Blanche Seller	granddaughter
			Frank Seller	grandson
SENN, John, Sr.	Pulaski	1839	Albert J. Senn	grandson
SHAFFSTALL, Christian	Steuben	1846	Carrie (Shaffstall) Kirlin	granddaughter
SHAUL, C. B.	Madison		Mrs. Clifford D. Shaul*	granddaughter-in-law
SHEDD, Kellogg	La Porte	1836	Milo Shedd	grandson
SHEETS, John	Wells	1837	Julia Decker	granddaughter
			Decker Lawrence	great-grandson
SHEETS, Mary	Marion		Clyde Rickes	grandson
SHEIDLER, George	Cass	1839	Oscar H. Hamburg	great-great-grandson
SHELHORN, John	Decatur	1820	Mabel (Shelhorn) Lines	granddaughter
			J. T. Shelhorn	great-grandson
SHERRILL, John	Lawrence	1830	Minnie (Sherrill) Rodgers	granddaughter
SHERRY, John	Tippecanoe	1833	Wilbur Clement	grandson
SHIPPEE, Nehimeah	La Porte	1835	Mrs. Francis Marion Wiltfong	great-granddaughter
SHIRK, Andrew	Franklin	1815	Lura (Chafee) Shirk*	great-granddaughter-in-law
SHIVELY, John	Madison	1847	Izora M. Shively	granddaughter
SHIVELY, Philip	Henry	1831	Salem L. Shively	grandson
SHOCKNEY, Charles	Randolph	1838	Nathan M. Shockney	grandson
SHOEMAKER, George	Boone	1835	Valeria Stark*	granddaughter-in-law
SHOEMAKER, Thomas	Morgan	1826	Mrs. Ernest Brown	great-great-granddaughter
SHORT, Aaron	Fountain	1837	Charles E. Short	great-grandson
SHORT, Thomas	LaGrange	1841	Lucile (Short) Schlosser	granddaughter
SHORT, Wesley	Lawrence	1818	Luther Short Ferguson	great-great-grandson
SHORTRIDGE, Samuel	Fayette	1825	Fred Hackleman	great-grandson
			DeWitt Sherwood	great-grandson
SHRADER, Aaron	Union	1831	Charles M. Shrader	grandson
SHUGART, John	Grant	1835	Arthur E. Shugart	great-grandson
SHUTER, Henry, Sr.	Dearborn	1845	Oliver E. Shuter	great-grandson
SHUTT, ——	Huntington	1840	David D. Shutt	great-grandson
SIDES, John	Gibson	1823	Mrs. William C. Hart	great-granddaughter
SIFTON, William	Decatur	1824	Mrs. Frank Martin	great-granddaughter
SIMMONS, Aaron	Randolph	1837	Eclar Simmons	grandson
SIMMONS, John B.	Hancock	1844	Mrs. Lloyd L. Apple	great-granddaughter
SIMPSON, James	Cass	1835	Robert L. Simpson	great-great-grandson
SIMPSON, Patrick	Knox	1783	Edward Jordan	not given
SINCLAIR, Isaac P.	Putnam	1836	Piercy L. Sinclair	great-grandson

It has a wonderful spring. My father Marion Hurst born in 1842 often drove a team of oxen here to drink. Buck & Jerry were the oxen.

Maude (Hurst) McNary
Putnam County

House Built in [18]28 still Stands. Bricks Burned on the farm has many interesting features. Windows High so Indians could not see in. Secret cuby holes for yarn etc. Barn built in [18]32. 3 Stories still main barn of the farm. [A]ll hand-hewn timbers.

Roy H. Meharry
Montgomery County

ANCESTOR	COUNTY	DATE	DESCENDANT	RELATIONSHIP
SIPPY, Jacob	Fulton	1843	Dessie (Oliver) Noyer	great-granddaughter
			Augustus Oliver	great-grandson
			Kenneth Oliver	great-grandson
SIX, John	Rush	1828	Russell Six	great-grandson
SIXBEY, Nicholas J.	LaGrange	1835	Lora E. (Sixbey) Klingaman	granddaughter
SKINNER, Daniel	Fayette	1818	W. A. Henry	great-grandson
SKINNER, Thomas	Cass	1837	Hugh Skinner	great-grandson
SMALL, James	Carroll	1833	Albert Garrett Small	grandson
SMELSER, Jacob	Wayne	1823	Walter J. Lafuse	great-grandson
SMITH, Aaron	Boone	1836/38	Homer E. Turpin	grandson
			Robert Turpin	great-grandson
SMITH, Benjamen	Hamilton	1837	Edith (Johnson) Wise	great-granddaughter
SMITH, Daniel	Owen	1826	Lois Wampler	great-granddaughter
SMITH, Daniel	Putnam	1838	Ernest R. Smith	grandson
SMITH, David	LaGrange	1832/43	Alice E. (Smith) Beisel	granddaughter
			Mrs. Clyde Smith*	granddaughter-in-law
			Hubert Smith	grandson
SMITH, Eppenetus	Franklin	1836	Grace (Eggleston) Abercrombie*	step-granddaughter
SMITH, Henry Coleman	Shelby	1849	David Smith	son
SMITH, Henry Edmond	Grant	1840	George Smith	grandson
SMITH, John	Tipton	1836	Leo D. Smith	grandson

KNOX COUNTY HISTORY, 1988

Steen family house in Knox County. See entry for Steen family on opposite page.

ANCESTOR	COUNTY	DATE	DESCENDANT	RELATIONSHIP
SMITH, Peter	Wayne	1816	Lydia B. Crowe*	granddaughter-in-law
SMITH, Peter	Wayne	1823	Frank C. Hayes	great-grandson
SMITH, Samuel	Hancock	1831	John H. Smith	grandson
SNIDER, Isaac	Hancock	1836	Otis C. Snider	grandson
SNIDER, John	Harrison	1809	Mrs. Richard Smoots	great-great-granddaughter
SNYDER, Michael	Union	1813	Azalea H. Johnson	great-granddaughter
			Joseph P. Snyder	great-great-grandson
SOAPER, Virgil	Posey	1837	Morton T. Elliott	grandson
SOLOWAY, James Rilla	La Porte	1847	Charles R. Walker	grandson
SONGER, Adam	Fountain	1837	Ruth (Songer) Henderson	great-granddaughter
SPAHR, James	Jay	1837	Oscar Spahr	grandson
SPAHR, John	Wayne	1812	Wayne Spahr	great-grandson
SPAHR, Philip	Jay	1837	Russel Spahr	grandson
SPENCER, Thomas	White	1834	Robert Spencer	grandson
SPITLER, Daniel	DeKalb	1833	Marie E. Wilson	great-great-great-granddaughter
SPRINGER, Jacob	Marion	1835	Eva (Springer) Williams	granddaughter
STAFFORD, James	DeKalb	1844	Nellie Farver	granddaughter
STAHL, Frederick	Jackson	1846	Edgar H. Stahl	great-grandson
STAIR, Joseph	Grant	1848	Glen Renbarger	grandson
STALCUP, Eli	Greene	1858	Lena May (Stalcup) Gimber	daughter
STALKER, David	Hamilton	1843	Ellis Barker	grandson
			Roxie Stalker	granddaughter
STANLEY, Evan	Grant	1820	Fred Stanley	grandson
STATON, George	Marion	1831	Laura (Stout) Aston	granddaughter
STAUFFER, Jacob	Wayne	1843	Edna C. Stauffer	granddaughter
			Roy H. Stauffer	grandson
STAYNER, Jessie	Steuben	1841	Albert W. Stayner	grandson
STEED, John F.	Jay	1845	Thomas I. Nixson	great-grandson
STEELE, Hans	Hancock		Paul Steele	great-grandson
STEELE, Samuel	Owen	1838	Cecil Ray Steele	great-grandson
			Ross Steele	great-grandson
STEEN, Ellender	Knox	1806	Catharine Dymple Robinson	granddaughter
			Charles Donahue Robinson	grandson
			Corrine Robinson	granddaughter
			Frances Aline Robinson	granddaughter
			Margaret Eleanor Robinson	granddaughter
			R. M. Robinson	grandson
STEINKAMP, Henry	Vanderburgh	1814	George J. Steinkamp	grandson

On March 5 John D. Brown bought 240 acres for $1.25 pr acre. . . . That John D. Brown was my husband's Grandfather. . . . No other owners [have] had title. Only 2 or 3 times in George Browns life time a mortgage was issued but was straightened out by either his mother or the wife. . . . [A]lways made 3 year rotation: corn wheat or oats then clover. My husband 3rd generation is 73 next month & was born here.

Mrs. James M. Brown
Howard County

A direct continuous line for 112 years. . . . The farm has never been owned or lived on by any one but the direct descendants of Jacob Van Huss since he purchased it in 1835.

Mrs. Enos Van Huss
Parke County

IHS INDIANA EXTENSION HOMEMAKERS COLLECTION

ANCESTOR	COUNTY	DATE	DESCENDANT	RELATIONSHIP
STEINMETZ, Fred	Vanderburgh	1842	Lyle O. Steinmetz	grandson
STELLE, Isaac	Fayette	1816	Mrs. Harry Robinson	great-granddaughter
STELZER, John	Jay	1839	Mrs. Fedalis Wagner	granddaughter
STEPHENS, Abednego	Benton	1841	J. Clark Griffith	great-great-grandson
STEPHENS, James	Owen	1836	Charles E. Kitch	great-grandson
STEVENSON, George	Jefferson	1827	Dale Stevenson	great-great-grandson
STEWART, Dixon	Vigo	1831	Eva Barnett	great-granddaughter
STEWART, James H.	Jackson	1853	Evangeline (Stewart) Booker	granddaughter
STEWART, Joseph	Montgomery	1829	Harman E. Stewart	great-grandson
STEWART, Samuel	Jackson	1853	Evangeline (Stewart) Booker°	great-grandniece
STIGLER, Samuel	Clay	1839	David Stigler	grandson
STINGER, Samuel	Rush	1833	Henry L. Stinger	great-grandson
STINSON, Arch	Fulton	1836	A. E. Stinson	grandson
STOCKTON, Newberry	White	1847	George W. Stockton	great-grandson
STOLTZ, Nicholas	Jay	1846	William H. Miller	grandson
			Mabel Norris	great-granddaughter
STONE, John	Carroll	1835	Lottie (Brown) Cooke	granddaughter
STONECIPHER, George	Harrison	1835	Ruth (Anderson) Brengman	great-great-granddaughter
STONER, David	Montgomery	1840	Paul Stoner	great-grandson
STOOPS, John Martin	Fayette	1837	William Stoops	grandson
STOOPS, William	Hamilton	1829	Sam N. Stoops	grandson

Milking cows

IHS INDIANA EXTENSION HOMEMAKERS COLLECTION

The deed was lost. . . . but a man planting the next spring found it. It is still readable and [the present owner] still has the buckskin grant in his possession.

Homer D. Elliott
Scott County

ANCESTOR	COUNTY	DATE	DESCENDANT	RELATIONSHIP
STORMONT, David	Gibson	1828	Anna M. Peoples	granddaughter
STORMONT, Robert	Gibson	1818	David L. Stormont°	grandnephew
STOUT, William	Whitley	1846	Orva (Stout) Young	granddaughter
STRANATHAN, William	Carroll	1835	Elizabeth (Stranathan) Askew	daughter
STRANGE, George	Grant	1841	Minnie Tudor	granddaughter
STRANGE, Hezekiah	Clinton	1829/31	Leodocia Strange	great-granddaughter
			Mace Strange	great-grandson
STRANGE, William Allen	Martin	1827	Pershing Jones	great-grandson
STROLE, ——	Vigo	1827	Joseph S. Strole	not given
STUART, Jehu, Jr.	Henry	1831	Laura (Stuart) Berg	great-granddaughter
			Lou V. Stuart	great-granddaughter
STUBBS, Joseph	Henry	1832	Charles L. Stubbs	grandson
STULTZ, John H.	Orange	1839	William A. Stultz	grandson
SULLIVAN, Henry	Jennings	1838	Edith Faye (Sullivan) Cooper	granddaughter
SUMMERS, Samuel	Warrick	1836	Ethel Anderson	great-granddaughter
			Helen Rives	great-granddaughter
			Hallie Youngblood	great-granddaughter
SUNMAN, John	Rush	1835	Everett Sunman	great-grandson
SWAIN, Elihu	Union	1832	Everett Higgins	great-grandson
SWAIN, Jonathan	Union	1835	Hollis D. Maxwell	grandson
SWAIN, Thomas	Rush		Clara E. Swain	granddaughter

ANCESTOR	COUNTY	DATE	DESCENDANT	RELATIONSHIP
SWAN, ——	Jefferson	1833	Ethel Gammon	not given
SWAN, Thomas	Jefferson	1838	Lee Swan	great-grandson
SWANK, Jacob	Huntington	1837	Charles Swank	grandson
SWEET, Benjamin	Morgan	1835	Edward M. Sweet	grandson
SWENEY, Isaac	Porter	1836	John Sweney	great-grandson
SWINFORD, John	Putnam	1825	Raymond Wright	great-grandson
SYMONS, Thomas	Hamilton	1836	Henry M. Symons	grandson
TALKINGTON, Jesse	Decatur	1838	Edgar Talkington	grandson
TALMAGE, Elisha	LaGrange	1837	Ernest Talmage	grandson
TAPPEN, Samuel	Union	1812	J. E. Tappen	great-grandson
TATE, William Augustus Hamilton	Fayette	1854	Mrs. Curtus Tate*	granddaughter-in-law
TAYLOR, Elias	Porter	1842	Judeth Rickes	granddaughter
TAYLOR, Filo	LaGrange	1836	Mrs. E. L. Grossman	granddaughter
TAYLOR, John	Johnson	1831	Horace Throckmorton	grandson
TERRELL, Richard	Clark	1789	Mrs. Oren T. Crum	great-great-great-great-great-granddaughte
TERRY, Robert	Ripley	1821	Mrs. O. H. Terry*	great-great-granddaughter-in-law
THACKERY, Joseph	Ripley	1833	Anna Dorrell	granddaughter
THARP, Jeremiah	Wayne	1823	Harold Tharp	great-grandson
THARP, John	Jay	1847	Hazel P. Black	granddaughter
THOM, Alex Craig	Jefferson	1830	Clay Thom	grandson

The old log part of our dwelling was put up when the land was entered in 1815. . . . We modernized, so to speak, but the old part of the house is fastened together with wooden pins, and nailed with square headed iron nails and two or three foot Poplar logs with "chinkin."

Roy H. Stauffer and
Edna C. Stauffer
Wayne County

Indiana Historical Society

CERTIFICATE OF RECOGNITIO

OF A

Centennial Farm Family

AWARDED TO

Thomas J. Williams

whose farm in the County of Benton a

the Township of Bolivar and Pine has remained

possession of the family for more than one hundred yea

Given at Indianapolis in January, 1949

PRESIDENT

ATTEST: Howard H. Peckham SECRETARY

INDIANA STATE ARCHIVES

The original of this Centennial Farm certificate is in the Williams folder, part of the Timmons family files in the Hoosier Homestead Project Collection at the Indiana State Archives. See the entries for the Timmons families on the opposite page.

Irwin K. Williams is shown standing in the old Timmons Cemetery, 9 July 1966. A note on the back of the photo states that Williams is the fifth generation descendant of original farm owner, Thomas J. Timmons. A copy of the photograph is in the Williams folder, part of the Timmons family files in the Hoosier Homestead Project Collection at the Indiana State Archives. See entries for Timmons families below.

INDIANA STATE ARCHIVES

Benj. Franklin Frazer owned & lived on site of Old Ft. Ouiatenon—First white settlement in Indiana. This farm is part of original Ft. site.

Lynette Frazer Kiser
Tippecanoe County

ANCESTOR	COUNTY	DATE	DESCENDANT	RELATIONSHIP
THOMAS, John	Wayne	1816	Clyde A. Thomas	great-great-grandson
THOMAS, John W.	Jefferson	1812	Daisy M. Thomas°	granddaughter-in-law
THOMAS, Owen	Vigo	1836	Rosetta Thomas°	granddaughter-in-law
THOMAS, William Winans	Fayette	1825	Hubard Trowbridge Thomas	grandson
THOMPSON, Alfred	Rush	1843	Ola Billman	grandson
THOMPSON, D. W.	Warrick	1840	Charles E. Thompson	son
THOMPSON, Jeremiah	Sullivan	1828	Jessie J. Bland	great-granddaughter
THOMPSON, Roger	Washington		Allen H. Thompson	great-great-grandson
THOMPSON, Samuel	Grant	1842	Eva T. Nesbitt	granddaughter
THOMPSON, Thomas	Rush	1843	Jessie A. (Thompson) Tillison	granddaughter
THORN, Stephen	Madison	1844	Claudia (Thorn) Carmony	great-granddaughter
THORNBURG, Joab	Randolph	1826	Leota Yoke	granddaughter
THORNBURG, Job	Randolph	1826	C. L. Thornburg	grandson
THORNBURGH, Henry	Fayette	1821	Eva Pearl (Thornburg) Caldwell	great-granddaughter
THORNTON, Job	Jefferson	1812	Grace Thornton°	great-granddaughter-in-law
THRELKELD, George	Boone	1834	James F. Reed	great-grandson
			Mary N. Reed	granddaughter
TILSON, Stephen	Johnson	1828	Rose Meredith	granddaughter
TIMMONS, Thomas	Benton	1829	Vivian (Timmons) McKinnis	great-granddaughter
TIMMONS, Thomas	Benton	1831	Laura Belle (Timmons) Williams	great-granddaughter
TINDALL, Isaac	Shelby	1830	William R. Tindall	great-great-grandson
TIPTON, Joshua	Fulton	1847	Bessie C. (Tipton) Gross	granddaughter

My great grandfather [Zachariah McAlister] and wife, Mary, with their first baby, fled to a fort a few miles south of S. Boston, from the Indians. Nothing to see [now] but a pile of rocks where the house stood. . . . My great grandfather was first settler at South Boston, Wash. Co.

Jesse Mahuron
Washington County

IHS INDIANA EXTENSION HOMEMAKERS COLLECTION

ANCESTOR	COUNTY	DATE	DESCENDANT	RELATIONSHIP
TODD, John	Dearborn	1834	Joseph G. Todd	grandson
TOWNSEND, Jesse	Wayne	1837	L. Jennie Townsend	great-granddaughter
TOWNSEND, Major	Johnson	1829	Roscoe E. Waltz	great-grandson
TRIPPET, Alexander	Gibson	1844	Aaron Trippet	grandson
			Eunice Trippet	granddaughter
TRIPPET, Waitman	Gibson	1814/30	Aaron Trippet	great-grandson
			Eunice Trippet	great-granddaughter
TROTTER, John	Harrison	1834	Kate Miller	granddaughter
TRUEBLOOD, Nathan	Washington	1817	Noble C. Trueblood	great-grandson
TRUEBLOOD, William N.	Washington		Noble C. Trueblood	grandson
TRUITT, Anderson	Clinton	1836	William A. Holloway	great-grandson
TUCKER, Elijah H.	Union	1835	Alpha Tucker	granddaughter
			George Tucker	grandson
TULL, Joseph	Shelby	1834	Mary (Smith) Gray	great-granddaughter
TULL, William	Jefferson	1820	Smyra/Smyrna F. Sever	grandson
TURNER, John	Monroe	1829	Turner Wiley	great-grandson
TWEEDY, James	Wabash	1847	Mrs. A. F. Tweedy*	daughter-in-law
TYNER, John	Fayette	1811	Walter Kolb	great-grandson
TYNER, William	Hancock	1833	Nellie Shadley	great-granddaughter
ULERY, John	Carroll	1839	Minnie Maxwell	great-granddaughter
ULREY, Sarah	Kosciusko	1837	Frank Aughinbaugh	great-great-grandson

ANCESTOR	COUNTY	DATE	DESCENDANT	RELATIONSHIP
UNGER, Jacob	Wabash	1840	Alvah Dubois	great-grandson
UTTERBACK, William	Johnson	1828	Harley O. Utterback	grandson
VANCE, Samuel	Henry		Orville Vance*	grandnephew
VANCICLES, Andrew	Marion	1829	Harold Earl Vansickle	great-great-grandson
VAN CLEAVE, Aaron	Montgomery	1828	Anna D. Van Cleave	great-great-granddaughter
			Cora Van Cleave	great-great-granddaughter
VAN CLEAVE, David	Montgomery	1831	Mrs. B. F. Van Cleave*	great-granddaughter-in-law
VANDIVIER, Madison	Johnson	1843	Otis M. Vandivier	grandson
VAN DOLAH, James	Allen	1837	Frances Bailey	granddaughter
VAN HUSS, Jacob	Parke	1835	Enos Van Hess	grandson
VANNICE, Peter C.	Hendricks	1831	Mrs. Lawrence G. Vannice*	grandniece-in-law
VAN PELT, Sutton	La Porte	1847	Sutton Van Pelt	grandson
VANSCHOICK, Hezekiah	Parke	1823	Susan Stokes	great-granddaughter
			W. R. Stokes	great-grandson
VANTREESE, George	Decatur	183?	John M. Vantreese	grandson
VEERKAMP, Mary	Decatur	1840	Clara (Veerkamp) Brancamp	great-great-granddaughter
			Lucy (Veerkamp) Parker	great-great-granddaughter
VERNON, Richard	Jefferson	1829	Victor Vernon	grandson
VINCENT, Thomas	Delaware	1837	A. Finley Vincent	great-grandson
VOELKER, Philip	Harrison	1836	Andrew Voelker	grandson
VOGLER, William	Bartholomew	1832	L. Marshall Vogler	grandson
WADE, Alfred	Switzerland		Charles Wade	grandson
WAGENER, Gad	Decatur	1837	Dorothy (Wagener) Petty	great-granddaughter
WAGGONER, Lot	Wabash	1848	C. M. Waggoner	son
WAGONER, Martin	Carroll	1830	Opal Wagoner	granddaughter
WAKINS, George	Montgomery	1831	Grace E. Peck	great-granddaughter
WALKER, Edward	Decatur	1840	Sarah Stella Walker	granddaughter
WALKER, Francis C.	Shelby	1833	Stella W. Gifford	great-granddaughter
			Lora Phillips	great-granddaughter
WALKER, Hugh	Delaware	1837	Atta (Walker) Shellabarger	granddaughter
WALLACE, John	Wayne	1814	Oliver T. Wallace	great-grandson
WALLACE, Robert	Grant	1847	C. N. Wallace	son
WALTER, Abraham	Steuben	1837	Enos Walter	grandson
WALTER, David	Spencer	1839	John Walter	great-grandson

John Sides, when he settled on this tract of land built his house on exactly the same spot where our present home now stands. My father and mother [Mr. and Mrs. Alva Carter] purchased this particular portion of the tract from my uncle who had received it from my grandfather. . . . I was born here and have never lived any place else. . . . I am the youngest great-grand-child of the original owner . . . and by odd coincidence I was born Sept. 22nd, the date of the first deed.

Mrs. William C. Hart
Gibson County

[Aaron Burget's] wife Elizabeth Hodkins Burget, carried their youngest child in her arms on horse-back from Brown Co. Ohio, to Johnson County Indiana when he was six weeks old sometime in April 182[?]. His name was Oliver Perry. Another of the sons Aaron born in Ohio in 1816, was the "deer hunter." . . . He was the hero of the Poem "Out on the Whetzel Trail."

Lewis B. Richardson
Johnson County

ANCESTOR	COUNTY	DATE	DESCENDANT	RELATIONSHIP
WALTZ, Joseph	Hamilton	1837	Carrie Bardonner	granddaughter
			Anna Waltz	granddaughter
			Charles Waltz	grandson
			Clara Waltz	grandson
			Dora Waltz	granddaughter
			Harry Waltz	grandson
			Samuel Waltz	son
			Walter Waltz	grandson
WANAMAKER, Christian	DeKalb	1841	Lucille Bearss	great-granddaughter
			Laura (Schramm) Rowley	great-granddaughter
			Charles R. Schramm	great-grandson
			Gerald C. Schramm	great-grandson
WANDEL, Stephen	Rush	1830	Leona C. Apple	great-granddaughter
WARE, James	Johnson	1829	Louisa Sanders	granddaughter
WARNER, Henry	Kosciusko	1837	Tom K. Warner	great-grandson
WASSMER, Wendle	Posey	1839	Albert Wassmer	grandson
			Joe Wassmer	grandson
			Lena Wassmer	granddaughter
			Tom Wassmer	grandson
WASSON, Joseph	Wayne	1805	Irene (Wasson) Berry	great-granddaughter
WATSON, Augustus	Warren	1831	Lewis V. Clem	grandson
WATSON, Eli	Whitley		Jacob B. Watson	grandson
WATSON, William	Harrison	1813	Mrs. Curry L. Miller	great-great-granddaughter
WEATHERHOLT, Jacob	Perry	1807	Arnold Leaf	great-great-grandson
			Kate (Leaf) Polk	great-granddaughter
WEBER, Gideon	Fulton	1843	Lyman Weber*	grandnephew
			Ralph Weber*	grandnephew
WEIR, John	Vigo	1831	Mildred A. Mays	granddaughter
WEIR, Norman	LaGrange		Lester Weir	grandson
WELCH, Samuel	Jefferson	1819	Finley Ralston	great-grandson
WELLMAN, Aaron	Rush	1831	Stella (Downey) Cofield	great-granddaughter
WERRY, Phillip	Warrick	1847	Charles H. Werry	great-great-grandson
WESSELS, Joseph	Decatur	1844	Joseph H. Schwegman*	great-grandnephew
WEST, Abram	Hendricks	1831	Homer West	great-grandson
WEST, Thomas	Union	1829	Clint Bryson	great-grandson
			Mrs. Fred Fisher	great-granddaughter
WHEAT, Daniel	Harrison	1847	Edward Wheat	great-grandson
			Frank Wheat Sr.	great-grandson
WHEATLEY, Joseph, Sr.	Marion	1830	Hattie B. Toon	great-granddaughter

ANCESTOR	COUNTY	DATE	DESCENDANT	RELATIONSHIP
WHITE, Peter	La Porte	1831	Belle Dolstrom	great-granddaughter
			George Dolstrom	great-grandson
WHITE, William	Vermillion		Charles Morey White	great-grandson
WHITEHEAD, Lazerus	La Porte	1837	Mary W. Peterson	granddaughter
WHITESELL, Jacob	Randolph	1836	Jacob E. Conklin	great-grandson
			Mary Jones	great-granddaughter
WHITESIDES, Lewis B.	Clark	1841	Mary Addie (Whitesides) Peeler	granddaughter
WHITTENBERGER, Jacob	Fulton	1840	Ina Brundige	daughter
WHITTENBERGER, William, Sr.	Fulton	1836	Odie Sausaman	great-grandson
WHITWORTH, William B.	Henry		Charles R. Whitworth	grandson
WICKER, Tally	Shelby	1828	Carroll L. Wicker	great-grandson
WIDEL, Lawrence	Ripley	1835	Edward Retzner	grandson
WIDNEY, Samuel	DeKalb	1836	Carl D. Carpenter	great-great-grandson
WILEY, James	Posey	1840	Fred D. Wiley	great-grandson
WILEY, Joseph	Jefferson	1820	Kathryn (Wiley) Kloepfer	great-granddaughter
WILEY, Robert F.	Grant	1847	Garr Brensford	grandson
WILGENBUSCH, Bernard	Dearborn	1844	William Wilgenbusch	son
WILHELM, John	Delaware	1851	Jesse E. Eiler	grandson
WILKERSON, John	Jefferson	1819	Ernest Wilkerson	grandson

Joseph Wasson, being a Revolutionary soldier—and having been wounded in a skirmish with the Tories in Carolina—was given a quarter section of land by the government. . . . This land lies about three and one-half miles east of Richmond, Indiana, along the Whitewater Valley.

Irene Wasson Berry
Wayne County

IHS INDIANA EXTENSION HOMEMAKERS COLLECTION

Samuel Summers rode horseback to Vincennes to buy this land. Which was entered from the government April 22-1836.

Mrs. Hallie Youngblood
Warrick County

IHS BASS PHOTO CO. COLLECTION, #29668

ANCESTOR	COUNTY	DATE	DESCENDANT	RELATIONSHIP
WILKINSON, William	Hamilton	1823	Mary Hunt	great-granddaughter
			Bertha Keller	great-granddaughter
			Sam Pursel	not given
			Howard Williamson	not given
WILLIAMS, Asa C.	Posey		Iva (Williams) Whitehead	granddaughter
WILLIAMS, Isaac	Lawrence	1816	Cornelia (Williams) Jones	granddaughter
WILLIAMS, John	Rush	1834	Ada (Williams) Stevens	granddaughter
WILLIAMS, John	Cass	1837	John Williams	grandson
WILLIAMS, Joseph	Posey	1835	Ida F. Redman	granddaughter
			Vincent Williams	great-grandson
			Virgil Williams Sr.	grandson
WILLIAMS, Thomas R.	Huntington	1846	Thomas R. Williams	grandson
			W. E. Williams	grandson
WILLIAMS, Welton	Knox	1836	Lester Williams	great-grandson
WILLIAMS, Wesley	Hancock	1837	John W. Simmons	grandson
WILLIAMS, William	Madison	1822	Ralph Williams	great-grandson
WILLIAMS, William	Posey	1836	Walter Williams	grandson
WILLIAMS, William Y.	Delaware	1844	Clinton W. Williams	great-grandson
WILLIAMSON, David	Cass	1836	Stewart C. Williamson	great-grandson
WILSON, Edward	Johnson	1833	Ruth A. Boner	great-granddaughter
WILSON, John	Washington	1821	Ezekiel L. Mead	grandson
WILSON, John R.	Vigo		Charles E. Wilson	grandson

ANCESTOR	COUNTY	DATE	DESCENDANT	RELATIONSHIP
WILSON, Joshua	Gibson	1812	George Wilson	grandson
			Henry Wilson	grandson
WILSON, Martha	Grant	1846	George Wilson	son
WILSON, Michael H. R.	Madison	1837	Evan W. Wilson	great-grandson
WILSON, Thomas	Posey	1830	Robert E. Wilson*	grandnephew
WIMER, William	Tipton	1843	Guy William Wimer	grandson
WINDLE, Augustine	Harrison	1818	Benton Windell	great-grandson
WINDLE, Benjamin	Harrison	1820	Dewey N. Hickman	great-grandson
WINDLE, Ephraim	Harrison	1832	John L. Windell	grandson
WINGARD, John	Carroll	1838	Clarence E. Wingard	grandson
			Louanna Wingard	granddaughter
			H. A. Wingard	grandson
WININGER, Samuel	Martin	1846	Cecil Ragsdale	grandson
			Allie Wininger*	daughter-in-law
			Merlin Wininger	grandson
WINSHIP, John	Rush	1833	Clif N. Winship	grandson
WISEMAN, Jacob	Washington	1837	Zella (Wiseman) Atkinson	great-granddaughter
WOEBKENBERG, Bernard, Sr.	Dubois	1840	Mrs. Bernard Woebkenberg Jr.*	granddaughter-in-law
WOLFE, Noah	Vigo	1845	Clara (Wolfe) Ramage	granddaughter
WOOD, Lorenzo D.	Decatur	1834	Mrs. Denzil Davis	granddaughter
WOOD, Robert	Henry	1827	Ray Morgan	grandson
WOOD, William H.	Randolph	1836	Norman L. Wood	grandson
WOODFIELD, Peter	Tippecanoe	1832	Asa Woodfield	grandson
			Pearl Woodfield	granddaughter
WOODS, Bartlett	Lake	1837	Sam B. Woods	son
WOODWARD, Jesse	Hendricks	1832	Katie (Woodward) Douglas	granddaughter
WOOLEY, Daniel	Ripley	1818	William Jesse Wooley	great-grandson
WOOLSEY, Temple	Pike	1839	Leota (Woolsey) Brewster	daughter
WORTH, Zeno	Vermillion	1822	Lee E. Porter	great-great-grandson
WRIGHT, Dan	Fayette	1840	Ray Stevens	great-grandson
WRIGHT, David	Jay	1835	Ada Wright*	granddaughter-in-law
WRIGHT, Edward	Randolph	1847	Lesta F. Curry	great-granddaughter
WRIGHT, Ephriam	Rush	1845	Luedith W. Simpson	granddaughter
WRIGHT, Henry	Wayne	1841	Ellen (Wright) Ranck	granddaughter
			Isaac Clayton Wright	grandson
			Lurena Wright	granddaughter

This farm carried a distinctive land mark until about fifteen years ago. Many places in southern Huntington Co. were described in usual conversation as being such and such direction from the Oak Tree. This oak tree stood in the crossroads 1 mile north of the center of Jefferson Township and was located on this Marshall-Pinkerton farm.

Mary A. Pinkerton
Huntington County

I was thinking of giving a sketch [about the house] to our N. Manchester News paper and leave a copy in a scrap book for our sons (Austin J. now 22 years of age and J. D. 22 mos. old) to read later in life as I know they will enjoy rehearsing it to folks around the Old homestead. . . . I am extremely anxious to receive this recognition for this farm because of our two sons who will carry on.

Freida Aughinbaugh
Kosciusko County

ANCESTOR	COUNTY	DATE	DESCENDANT	RELATIONSHIP
WRIGHT, Isaac B.	Spencer	1823	Amos P. Wright	grandson
			Shelby Markland Wright	great-grandson
			Amos P. Wright II	great-great-grandson
WYKOFF, Garrett	Rush	1846	Luther Nixon	great-grandson
WYSONG, David	Randolph	1838	John H. Wysong	great-grandson
YANTIS, Aaron	Cass	1841	Emma Yantis	granddaughter
YEAGER, Laudon P.	Carroll	1843	Jesse L. Yeager	grandson
YERKES/JERKES, Josiah	Carroll	1835/37	Cloyd S. Yerkes	great-grandson
			Iva I. Yerkes*	great-granddaughter-in-law
YOUNT, Daniel	White	1845	Carl Clouse*	grandnephew
ZAEHNLE, Arbogast	St. Joseph	1837	Dorothy D. Akers	great-granddaughter
ZINKAN, ——	Daviess	1841	Quintilla Zinkan	granddaughter

Bysspho

Index to Descendants

Compiled by

Ruth Dorrel

Of course all the old buildings are gone but I can remember most of them as I am past 66 years old. . . . Grandfather bought this place for the water he has told me. There is a cave with spring in same on place Cecil [Peacher] has. Also another cave that has an apple storage room that holds about 10,000 bushels of apples. These apples are let in and out with windlass and motor on narrow guage [*sic*] track.

Harry Peacher
Orange County

Valentine Barnett . . . lived at what is now Rich Valley on the banks of the Erie Canal. Two years prior to this purchase, my grandfather, Michael Dice came in on a canal boat, got a job with Barnett, and married Elizabeth Barnett. Evidently this purchase 2½ miles from his home in Rich Valley was made in order to set young Michael and his bride in a home, for on July 12th 1847 [Valentine] Barnett deeded the 2 80-acre lots to Michael Dice. Consideration $300.00.

Claude A. Dice
Miami County

IHS BASS PHOTO CO. COLLECTION, #63504

Orchard, 1928

Jacob Weatherholt received his part of a track of land for services in the Revolutionary War.

Kate Leaf Polk
Perry County

"This barn [in background of photo] was first hip roof barn in this part of county. Early log house was in the same place as this one. Electric line was first high power line in this part of state. Road 40 years ago was paved with logs. Looking N. W. across Indiana Highway #1. John Philip and dog Butch in foreground (John David Carpenter in rear). Jan. 1947." (Carl D. Carpenter, applicant for Harvester Farms, DeKalb County)

[James McCord and Hannah (Morris) McCord] lived on a farm about a mile north of Independence, [and my great-grandfather] walked from there to where he Homesteaded night and morning while building the house and when they raised the barn my Great-Grandmother baked pies in a brick oven at her neighbors about a mile away. There were as many as 100 Hard Maple trees north of the house the residence was in the center of the farm.

Hannah E. Anderson
Warren County

Jacob Pence was the first settler in Sims township which was named after his wife's uncle Joseph Sims. . . . The [first tax receipt] was for $1.93, dated 1846.

Benjamin F. Pence
Grant County

IHS INDIANA EXTENSION HOMEMAKERS COLLECTION

Farm was purchased from a soldier, who had a grant for it from the war of 1812.

Jesse Charles Andrew
Tippecanoe County

In 1832 Samuel Holmes came from Montgomery Co. Ohio and "purchased" land known as "canal land" (set aside to pay for Wabash Erie Canal) for $1.25 per acre. . . . [The present owners] have two sons 12 and 9 years to grow to care for it. The lovely large brick home is well preserved and stands as a tribute to the sixth generation. [This homestead was known as Leatherwood Farms.]

Fay Buckley Wise
Carroll County

IHS INDIANA EXTENSION HOMEMAKERS COLLECTION

[The house and barn] are both hewed frame buildings. The house is furnished in walnut which grew on the farm, and both are still in use altho remodeled and improved. My mother Lucinda Macy was born and reared in this house and on her marriage to Abel Gilbert in 1866, they took over the operation of the farm, living in the house with, and caring for her parents, James and Anna Macy, the remainder of their lifetime.

Henry Wilson Gilbert
Henry County

Elisha Chambers entered land [1818]. . . . He was my Great Great Grand Father. He and his father James Chambers came from Pennsylvania or Maryland to Burk[e] Co. North Carolina, to Owen, Co. . . . At one time on this farm there were one general store, one harness shop, two blacksmith shops, and several houses. It was called Chambersville. The buildings are all gone except the four room house I live in and two rooms of the house that was used for the store I now use for a feed room.

Ivan Chambers
Owen County

These two brothers [Conrad and Justus Minnick] with their families emigrated to America in about 1837 and settled in Somerset Co. Penn[sylvania]. . . . In 1844 or 45 they moved to Tipton township, Cass Co. Indiana, just a few years before the town of Walton was settled in 1852. . . . Justice Minnick my Great Grand Father died in 1847 and is buried out in the field on this farm. He also had a son named Justice and his brother had a son named Justice, which makes it rather confusing.

Alvin E. Minnick
Cass County

Robert [Harrison] at the age of 18 came from Butler Co. Ohio in 1825 & entered a section of land @ $1.25 an acre. The family moved in 1827. They worked with an ox team. Robert aided in clearing ground for first building at Wabash College in Crawfordsville.

Henry A. Harrison
Montgomery County

Index to Counties

Compiled by

Ruth Dorrel